99-WWF-5

EVALUATING AND OPTIMIZING SOURCE TREATMENT TECHNOLOGIES TO IMPROVE THE BIODEGRADABILITY OF ORGANIC COMPOUNDS

by:

Mirat D. Gurol
Ahmet Aydogan
Badri Badriyha
Fatih Buyuksonmez
Alex Christensen
Tuba Ertas
Temesgen Garoma
Milica Kaludjerski
Greg Loraine
Melissa Valdovinos

San Diego State University
Department of Civil and Environmental Engineering

2004

WERF

The Water Environment Research Foundation, a not-for-profit organization, funds and manages water quality research for its subscribers through a diverse public-private partnership between municipal utilities, corporations, academia, industry, and the federal government. WERF subscribers include municipal and regional water and wastewater utilities, industrial corporations, environmental engineering firms, and others that share a commitment to cost-effective water quality solutions. WERF is dedicated to advancing science and technology addressing water quality issues as they impact water resources, the atmosphere, the lands, and quality of life.

For more information, contact:
Water Environment Research Foundation
635 Slaters Lane, Suite 300
Alexandria, VA 22314-1177
Tel: (703) 684-2470
Fax: (703) 299-0742
www.werf.org
werf@werf.org

This report was co-published by the following organizations. For nonsubscriber sales information, contact:

IWA Publishing
Alliance House, 12 Caxton Street
London SW1H 0QS, United Kingdom
Tel: +44 (0) 20 7654 5500
Fax: +44 (0) 20 7654 5555
www.iwapublishing.com
publications@iwap.co.uk

Library of Congress Catalog Card Number: 2004106251

IWAP ISBN: 1-84339-702-1 WERF ISBN: 1-893664-87-2

San Diego State University

ACKNOWLEDGMENTS

The Principal Investigator gratefully acknowledges the help received from each individual listed below that led to successful completion of this project.

Report Preparation:

Principal Investigator:

Mirat D. Gurol, Ph.D.

Project Team

Ahmet Aydogan
Badri Badriyha, Ph.D., P.Eng.
Fatih Buyuksonmez, Ph.D., P.Eng.
Alex Christensen
Tuba Ertas, Ph.D.
Temesgen Garoma
Barbara Hemmingsen, Ph.D.
Milica Kaludjerski
Greg Loraine, Ph.D., P.Eng.
Ismael Maya
Melissa Valdovinos

Project Subcommittee

Robert Cowan, Ph.D.
Rutgers University

Joe Jacangelo, Ph.D.
Montgomery Watson

Andrew Middleton, Ph.D., P.Eng., D.E.E.
Corporate Environmental Solutions LLC

Richard Poduska, Ph.D.
Eastman Kodak

Robert Reich, P.Eng.
DuPont Company

Water Environment Research Foundation Staff

Director of Research: Daniel M. Woltering, Ph.D.
Project Manager: Lola Olabode, M.P.H.

ABSTRACT AND BENEFITS

Abstract:

The main objective of this research was to investigate the capabilities of three chemical oxidation processes as pretreatment technologies with the goal of making wastewaters containing persistent organic compounds amenable to biotreatment. The processes investigated are ozonation, ultraviolet radiation/ hydrogen peroxide (UV/H_2O_2) and Fenton reaction. The studies were focused on two organic compounds: dichloro diethyl ether (DCDE) and methyl tertiary butyl ether (MTBE). Synthetic solutions, real groundwater, and industrial wastewater were used for experimentation.

The experimental method involved oxidation of solutions of the target organic compounds at various percentages by the three oxidation processes. The pre-oxidized solutions of the organic compounds were subjected to biodegradation and toxicity studies. Four different respirometric tests (two short-term, one mid-term, and one long-term) and a bench scale sequencing batch reactor (SBR) test were used for full assessment of the effectiveness of the chemical oxidation processes. Activated sludge, as acclimated and non-acclimated to the organic compounds, was used as the test culture.

All three chemical oxidation processes were able to transform DCDE and MTBE to a variety of oxidation by-products. Oxidation byproducts by all three oxidation methods were significantly more biodegradable than the nonoxidized parent compounds. This project clearly showed that integration of chemical oxidation processes with biotreatment may lead to effective handling of "problem" wastewaters.

Benefits:

Many synthetic organic compounds have been found to be recalcitrant in municipal or industrial wastewater biotreatment systems. A substantial number of classes of these compounds have been found to be oxidized by chemical oxidation processes, such as ozonation, UV radiation/hydrogen peroxide and Fenton treatment. However, total mineralization in such processes might be extremely costly. Controlled partial pretreatment in such processes followed by biotreatment may be a more cost-effective approach for dealing with some classes of persistent compounds.

- Chemical oxidation can be used as a pretreatment technique to oxidize persistent compounds, such as dichloro diethyl ether (DCDE) and methyl tertiary butyl ether (MTBE) in industrial wastewaters.

- Ozonation, ultraviolet radiation/ hydrogen peroxide (UV/H_2O_2) and Fenton treatment are all effective in eliminating DCDE and MTBE, although ozonation is less effective than the latter two processes.

- Ozonation, UV/H_2O_2 and Fenton treatment convert non-biodegradable parent compounds into biodegradable products.

- The biodegradability and the toxicity of oxidation by-products to activated sludge can be evaluated by the experimental protocol developed in this study.

- Chemical oxidation followed by biological degradation represents an effective integrated treatment system for many industrial wastes that contain toxic or non-biodegradable compounds.

Keywords: Chemical oxidation, biotreatment, ozone, ultraviolet radiation/ hydrogen peroxide, Fenton reaction, dichloro diethyl ether, methyl tertiary butyl ether, pretreatment.

TABLE OF CONTENTS

Acknowledgments iii
Abstract and Benefits iv
List of Tables ix
List of Figures xi
List of Acronyms xv
Executive Summary ES-1

1.0 Introduction 1-1
1.1 Rationale for Research 1-1
1.2 Background 1-2
1.2.1 Chemical Oxidation Processes 1-2
1.2.2 Use of Chemical Oxidation for Pretreatment Prior to Biotreatment 1-8
1.2.3 Test Compounds 1-12

2.0 Objectives and Experimental Approach 2-1
2.1 Objectives 2-1
2.2 Experimental Methods 2-1
2.2.1 General Approach 2-1
2.2.2 Chemical Oxidation Tests 2-3
2.2.3 Activated Sludge Culture 2-5
2.2.4 Biodegradability Tests 2-6
2.2.5 Inhibition Tests: Test Methodology 2-11
2.2.6 SBR Tests 2-12
2.2.7 Analytical Methods 2-14
2.2.8 Chemicals 2-15

3.0 Chemical Oxidation Tests 3-1
3.1 Ozonation 3-1
3.1.1 Characterization of Ozonation System 3-1
3.1.2 Ozonation of DCDE 3-1
3.1.3 Ozonation of MTBE 3-5
3.2 Fenton Treatment 3-7
3.2.1 Fenton Treatment of DCDE 3-7
3.2.2 Fenton Treatment of MTBE 3-9
3.3 UV/H_2O_2 Treatment 3-11
3.3.1 UV/H_2O_2 Treatment of DCDE 3-11
3.3.2 UV/H_2O_2 Treatment of MTBE 3-14
3.4 Oxidation By-Products of DCDE 3-15
3.5 Oxidation By-Products of MTBE 3-21

4.0 Assessment of Biodegradation for DCDE 4-1
4.1 Short-term Biodegradation Tests 4-1
4.1.1 Ozone-Treated DCDE 4-2
4.1.2 Fenton-Treated DCDE 4-3
4.1.3 UV/H_2O_2-Treated DCDE 4-5
4.1.4 Comparison of Treatments 4-6
4.2 Long-Term Biodegradation Tests 4-7
4.2.1 Ozone-Treated DCDE 4-7
4.2.2 Fenton-Treated DCDE 4-8
4.2.3 UV/H_2O_2-Treated DCDE 4-8
4.2.4 Comparison of Treatments 4-8
4.3 Mid-Term Biodegradation Tests 4-18
4.3.1 Ozone-Treated DCDE 4-18
4.3.2 Fenton-Treated DCDE 4-19
4.3.3 UV/H_2O_2-Treated DCDE 4-20
4.3.4 Comparison of Treatments 4-20

5.0 Assessment of Biodegradation for MTBE 5-1
5.1 Short-Term Biodegradation Tests 5-1
5.1.1 Ozone-Treated MTBE 5-1
5.1.2 Fenton-Treated MTBE 5-2
5.1.3 UV/H_2O_2-Treated MTBE 5-3
5.1.4 Comparison of Treatments 5-3
5.2 Long-Term Biodegradation Tests 5-4
5.2.1 Ozone-Treated MTBE 5-4
5.2.2 Fenton-Treated MTBE 5-5
5.2.3 UV/H_2O_2-Treated MTBE 5-10
5.2.4 Comparison of Treatments 5-10
5.3 Mid-Term Biodegradation Tests on UV/H_2O_2-Treated MTBE 5-15

6.0 Toxicity Assessment 6-1
6.1 Tests with DCDE 6-1
6.1.1 Ozone-Treated DCDE 6-1
6.1.2 Fenton-Treated DCDE 6-2
6.1.3 UV/H_2O_2-Treated DCDE 6-3
6.1.4 Comparison of Treatments for DCDE 6-4
6.2 Tests with MTBE 6-6
6.2.1 Ozone-Treated MTBE 6-7
6.2.2 Fenton-Treated MTBE 6-8
6.2.3 UV/H_2O_2-Treated MTBE 6-9
6.2.4 Comparison of Treatments for MTBE 6-9

7.0 Sequencing Batch Reactor (SBR) Tests 7-1
7.1 Assessment of Biodegradation 7-1
7.2 Tests with DCDE 7-2
7.2.1 Ozone-Treated DCDE 7-3
7.2.2 Fenton-Treated DCDE 7-8

7.2.3 UV/H_2O_2-Treated DCDE 7-10
7.2.4 Comparison of Treatments for DCDE 7-15
7.2.5 SBR Tests with Non-Oxidized DCDE 7-16
7.3 Tests with MTBE 7-18

8.0 Experiments with Groundwater and Wastewater 8-1
8.1 MTBE Containing Groundwater 8-1
8.1.1 Characteristics and Chemical Oxidation 8-1
8.1.2 Long-Term Respirometric Tests 8-2
8.1.3 SBR Tests on Groundwater Samples 8-4
8.2 Hollow Fiber Membrane Waste 8-6
8.2.1 Waste Characterization and Chemical Oxidation 8-6
8.2.2 Long-Term Respirometric Tests 8-10

9.0 Conclusions and Recommendations 9-1
9.1 Conclusions 9-1
9.1.1 Chemical Oxidation 9-1
9.2 Assessment of Biodegradation for Pre-oxidized DCDE 9-2
9.3 Assessment of Biodegradation for Pre-oxidized MTBE 9-3
9.4 Assessment of Biodegradation for Pre-oxidized Groundwater and Industrial Wastewater 9-4
9.5 Evaluation of the Biodegradability and Toxicity Tests 9-4
9.6 Recommendations 9-6

Appendix A A-1
Appendix B B-1

References R-1

LIST OF TABLES

1-1 Comparison of Chemical Oxidation Processes ... 1-5
1-2 Properties of the Test Compounds ... 1-13
2-1 Syntho Solution Constituents ... 2-13
2-2 List of Chemicals Used in this Research ... 2-16
3-1 Fenton Treatment of DCDE for Various Reagent Doses ... 3-8
3-2 Comparison of Results with Ferric Sulfate and Ferric Chloride ... 3-9
3-3 Comparison of Intermediate Products for All Oxidation Methods (DCDE Removal: 61-78%) ... 3-18
3-4 Comparison of Intermediate Products for All Oxidation Methods (DCDE Removal: 85–90%) ... 3-18
3-5 DCDE Oxidation Products Identified by Mass Spectrometry and IC ... 3-21
3-6 Comparison of Intermediate Products from MTBE for Three Oxidation Methods (MTBE removal: 80-90%) Product Concentrations in mg/L ... 3-26
3-7 Comparison of Intermediate Products from MTBE for Three Oxidation Methods (MTBE removal: 80-90%) Normalized to Moles per 100 moles to MTBE Removed.. 3-26
4-1 Biodegradation of 94% Oxidized DCDE by Ozone. Non-Acclimated Culture ... 4-18
4-2 Biodegradation of 94% Oxidized DCDE by Ozone. Acclimated Culture ... 4-18
4-3 Biodegradation of 90% Oxidized DCDE by Fenton Reagent Non-Acclimated Culture ... 4-19
4-4 Biodegradation of 90% Oxidized DCDE by Fenton Reagent Acclimated Culture ... 4-19
4-5 Biodegradation of 100% Oxidized DCDE by UV/H_2O_2 Non-Acclimated Culture ... 4-20
4-6 Biodegradation of 100% Oxidized DCDE by UV/H_2O_2 Acclimated Culture ... 4-20
4-7 Comparison of Treatments on the Basis of Overall Removal of Organic Matter (npCOD and npTOC) in Terms of Average Values and Range of Data of Duplicate Tests during 6 Hours of Biodegradation Experiments ... 4-21
5-1 Biodegradation of 98% Oxidized MTBE by UV/H_2O_2 Acclimated Culture ... 5-15
5-2 Biodegradation of 98% Oxidized MTBE by UV/H_2O_2 Non-Acclimated Culture ... 5-15
6-1 Inhibition of Respiration of Acclimated Culture Due to Non-Oxidized MTBE ... 6-7
6-2 Inhibition of Respiration of Non-Acclimated Culture Due to Non-Oxidized MTBE ... 6-7
6-3 Inhibition of Respiration of Acclimated Culture Due to Ozonated MTBE ... 6-7
6-4 Inhibition of Respiration of Non-Acclimated Culture Due to Ozonated MTBE ... 6-8
6-5 Inhibition of Respiration of Acclimated Culture Due to Fenton-Treated MTBE ... 6-8
6-6 Inhibition of Respiration of Non-Acclimated Culture Due to Fenton-Treated MTBE ... 6-8
6-7 Inhibition of Respiration of Acclimated Culture Due to UV/H_2O_2-Treated MTBE ... 6-9
6-8 Inhibition of Respiration of Non-Acclimated Culture Due to UV/H_2O_2-treated MTBE. 6-9
7-1 Target COD and TOC Contributions of Oxidized DCDE to the Feed Solution ... 7-3
7-2 Estimation of Percent Biodegradation of DCDE By-products ... 7-5
7-3 Estimation of Percent Biodegradation of Ozonation By-Products of DCDE (Experiment 7) ... 7-7
7-4 Estimation of Percent Biodegradation of Fenton By-Products of DCDE (Experiment 6) ... 7-10
7-5 Estimation of Percent Biodegradation of UV/H_2O_2 Oxidation By products of DCDE (Experiment 1) ... 7-13

7-6 Estimation of Percent Biodegradation of UV/H_2O_2 Oxidation By-Products of DCDE (Experiment 2) ... 7-15
7-7 Comparison of Biodegradation Based on TOC Reduction for DCDE Utilizing SBRs. 7-16
7-8 Target COD and TOC Contributions of Oxidized MTBE to the Feed Solution ... 7-18
7-9 Estimation of Percent Biodegradation of UV/H_2O_2 Oxidation By-Products of MTBE (Experiments 10 and 11) ... 7-21
7-10 Estimation of Percent Biodegradation of UV/H_2O_2 Oxidation By-Products of MTBE (Experiment 12) ... 7-23
7-11 Comparison of Biodegradation Based on TOC Reduction for MTBE Utilizing SBRs. 7-23
8-1 Groundwater Characteristics ... 8-1
8-2 Change in TOC Values During Respirometric Experiment ... 8-2
8-3 Makeup of Synthetic Waste ... 8-7
8-4 UV/H_2O_2 Oxidation of PVP ... 8-8
8-5 Reaction Rates with Hydroxyl Radicals ... 8-9

A-1 The Originally Proposed Test Chemicals and Their Selected Properties ... A-1

LIST OF FIGURES

2-1 Experimental Set-up for Ozonation 2-3
2-2 UV/H_2O_2 Oxidation Experimental Setup 2-5
2-3 Respiration Rates of Microorganisms with Different Dilutions of DCDE After 75% Oxidation by Ozone 2-7
2-4 Schematic of Experimental Set-up for Mid-term Biodegradation Tests 2-10
2-5 Experimental Set-up for SBR Studies 2-12
3-1 Influent and Effluent Ozone Gas Concentration at pH 3 3-2
3-2 Accumulation of Ozone in Solution at pH 3 3-2
3-3 Determination of the Mass Transfer Coefficient, k_La 3-3
3-4 Oxidation of DCDE by Ozone 3-3
3-5 Influent and Effluent Ozone Gas Concentrations During DCDE Oxidation 3-4
3-6 Ozone Concentration in Water During DCDE Oxidation 3-5
3-7 Ozonation of DCDE with and without TBA 3-5
3-8 Removal of MTBE by Stripping During Oxygenation 3-6
3-9 Removal of MTBE by Ozonation and Stripping 3-6
3-10 Ozone Influent and Effluent Gas Concentration During MTBE Ozonation 3-7
3-11 Ozone Concentration in Liquid During MTBE Ozonation 3-7
3-12 MTBE Removal by Stepwise Additions of Fe(III) H_2O_2 3-9
3-13 Comparison of DCDE and MTBE Removal by Stepwise Addition of Fe(III)/H_2O_2 3-10
3-14 MTBE Removal with Stepwise Fe(II)/H_2O_2 Additions 3-10
3-15 MTBE Removal During Fenton Treatment by Chemical Oxidation and Physical Removal 3-11
3-16 Removal of DCDE by UV Photolysis 3-12
3-17 Removal of DCDE by UV/H_2O_2 3-13
3-18 H_2O_2 Remaining after DCDE Removal 3-13
3-19 DCDE Removal by UV/H_2O_2 by Stepwise H_2O_2 Addition 3-14
3-20 MTBE Removal by UV/H_2O_2 by Stepwise H_2O_2 Addition 3-14
3-21 Oxidation By-Products of DCDE by Ozone 3-15
3-22 Oxidation By-Products of DCDE of Fenton Treatment by $FeCl_3$ 3-16
3-23 Oxidation By-Products of DCDE of Fenton Treatment by $Fe_2(SO_4)$ 3-16
3-24 Oxidation By-Products of DCDE by UV/H_2O_2 3-17
3-25 Dechlorination of DCDE During Its Oxidation 3-19
3-26 Major Intermediate Products of Ozonation of MTBE in mg/L 3-22
3-27 Major Intermediate Products of Ozonation of MTBE in mM C 3-23
3-28 Major Intermediate Products of Fenton Treatment of MTBE in mg/L 3-23
3-29 Major Intermediate Products of Fenton Treatment of MTBE in mM C 3-24
3-30 Major Intermediate Products of UV/H_2O_2 of MTBE in mg/L 3-25
3-31 Major Intermediate Products of UV/H_2O_2 of MTBE in mM C 3-25
4-1 Respiration Rates of Microorganisms with Different Dilutions of DCDE After 25% of Oxidation by Ozone 4-1
4-2 Respiration Rates of Non-Acclimated Cultures for Ozone-Treated DCDE Solutions 4-2
4-3 Respiration Rates of Acclimated Cultures for Ozone-Treated DCDE Solutions 4-3
4-4 Respiration Rates of Non-Acclimated Cultures for Fenton-Treated DCDE 4-4

4-5 Respiration Rates of Acclimated Cultures for Fenton-Treated DCDE 4-4
4-6 Respiration Rates of Non-Acclimated Cultures for UV/H_2O_2-Treated DCDE 4-5
4-7 Respiration Rates of Acclimated Cultures for UV/H_2O_2-Treated DCDE 4-5
4-8 Comparison of μ_{max} for Different Oxidation Methods and Acclimated and Non-Acclimated Cultures 4-6
4-9 Comparison of K_s for Different Oxidation Methods and Acclimated and Non-Acclimated Cultures 4-7
4-10 Oxygen Decline and Carbon Dioxide Production with Non-Acclimated Culture of Ozone-Treated DCDE 4-9
4-11 Oxygen Decline and Carbon Dioxide Production with Acclimated Culture of Ozone-Treated DCDE 4-10
4-12 Oxygen Decline and Carbon Dioxide Production with Non-Acclimated Culture of Fenton-Treated DCDE 4-11
4-13 Oxygen Decline and Carbon Dioxide Production with Acclimated Culture of Fenton-Treated DCDE 4-12
4-14 Oxygen Decline and Carbon Dioxide Production with Non-Acclimated Culture of UV/H_2O_2-Treated DCDE 4-13
4-15 Oxygen Decline and Carbon Dioxide Production with Acclimated Culture of UV/H_2O_2-Treated DCDE 4-14
4-16 Comparison in Net Oxygen Decline and Carbon Dioxide Production of Fenton, Ozone, and UV/H_2O_2-Treated Samples Using Acclimated and Non-Acclimated Cultures 4-15
4-17 Comparison in Net Oxygen Decline and Carbon Dioxide Production of Fenton, Ozone and UV/H_2O_2-Treated Samples Using Acclimated and Non-Acclimated Cultures During the First 24 Hours 4-17
5-1 Biodegradability of Ozone-Treated MTBE with (a) Acclimated and (b) Non-Acclimated Cultures 5-1
5-2 Biodegradability of Fenton-Treated MTBE with (a) Acclimated and (b) Non-Acclimated Cultures 5-2
5-3 Biodegradability of UV/H_2O_2-treated MTBE by (a) Acclimated and (b) Non-Acclimated Cultures 5-3
5-4 Comparison of μ_{max} after Ozone, Fenton, or UV/H_2O_2 Treatment of MTBE 5-4
5-5 Net Oxygen Decline and Carbon Dioxide Production with Acclimated Culture of Ozone-Treated MTBE 5-6
5-6 Net Oxygen Decline and Carbon Dioxide Production with Non-acclimated Culture of Ozone-Treated MTBE 5-7
5-7 Net Oxygen Decline and Carbon Dioxide Production with Acclimated Culture of Fenton-Treated MTBE 5-8
5-8 Net Oxygen Decline and Carbon Dioxide Production with Non-acclimated Culture of Fenton-Treated MTBE 5-9
5-9 Net Oxygen Decline and Carbon Dioxide Production with Acclimated Culture of UV/H_2O_2-Treated MTBE 5-11
5-10 Net Oxygen Decline and Carbon Dioxide Production with Non-acclimated Culture of UV/H_2O_2-Treated MTBE 5-12
5-11 Comparison of Net O_2 Decline and CO_2 Production of Ozone, Fenton, and UV/H_2O_2-Treated Samples Using Acclimated and Non-Acclimated Cultures at 75% Oxidation Level 5-13

5-12 Comparison of Net O_2 Consumption and CO_2 Production of Fenton, Ozone, and UV/H_2O_2-Treated Samples Using Acclimated and Non-Acclimated Cultures at 100% Oxidation Level 5-14
5-13 Comparison of Samples and Control Experiments for TOC Removal 5-16
6-1 Exogenous Inhibition of Cultures in the Presence of DCDE 6-2
6-2 Exogenous Inhibition of Cultures Exposed to DCDE and Its Oxidation By-Products After Ozonation 6-2
6-3 Exogenous Inhibition of Cultures Exposed to DCDE and Its Oxidation By-Products After Fenton Treatment 6-3
6-4 Exogenous Inhibition of Cultures Exposed to DCDE and Oxidation By-Products After UV/H_2O_2 Treatment 6-4
6-5 Comparison of Inhibition on Acclimated Culture of Oxidized DCDE 6-5
6.6 Comparison of Inhibition on Non-Acclimated Culture of Oxidized DCDE 6-5
6-7 Comparison of Inhibition on Acclimated Cultures for Oxidized and Non-Oxidized DCDE 6-6
6-8 Comparison of Inhibition on Non-Acclimated Cultures for Oxidized and Non-Oxidized DCDE 6-6
6-9 Comparison of Percent Inhibition for Oxidized MTBE by the Three Oxidation Methods 6-10
7-1 A Graphic Example of Estimation of the Biodegradable and Nonbiodegradable Fractions of the Feed Solutions 7-2
7-2 Average Suspended Solids vs. Time (Experiment 7) 7-4
7-3 Average COD Removal Over Time (Experiment 7) 7-4
7-4 Average TOC Removal Over Time (Experiment 7) 7-5
7-5 Average Suspended Solids vs. Time (Experiment 8) 7-6
7-6 Average COD Removal vs. Time (Experiment 8) 7-6
7-7 Average TOC Removal vs. time (Experiment 8) 7-7
7-8 Average Suspended Solids vs. Time (Experiments 3-6) 7-9
7-9 Average COD Removal Over Time (Experiments 3-6) 7-9
7-10 Average TOC Removal Over Time (Experiments 3-6) 7-10
7-11 Average Suspended Solids vs. Time (Experiment 1) 7-11
7-12 Average COD Removal Over Time (Experiment 1) 7-12
7-13 Average TOC Removal Over Time (Experiment 1) 7-12
7-14 Average Suspended Solids vs. Time (Experiment 2) 7-14
7-15 Average COD Removal Over Time (Experiment 2) 7-14
7-16 Average TOC Removal Over Time (Experiment 2) 7-15
7-17 COD Removal in DCDE-Fed Reactors 7-17
7-18 TOC Removal in DCDE-Fed Reactors 7-17
7-19 DCDE Reduction vs. Time 7-18
7-20 Average Suspended Solids vs. Time (Experiments 10 and 11) 7-19
7-21 Average COD Removal Over Time (Experiments 10 and 11) 7-20
7-22 Average TOC Removal Over Time (Experiments 10 and 11) 7-20
7-23 Average Suspended Solids vs. Time (Experiment 12) 7-21
7-24 Average COD Removal (Experiment 12) 7-22
7-25 Average TOC Removal (Experiment 12) 7-22
8-1 Change in MTBE Concentration with Time During Groundwater Oxidation with H_2O_2/UV Technique 8-2

8-2 Net Oxygen Decline for Respirometric Experiments with Groundwater 8-3
8-3 Net CO_2 Production Results for Respirometric Experiments with Groundwater 8-4
8-4 Suspended Solids vs. Time 8-5
8-5 COD Removal Over Time 8-5
8-6 TOC Removal Over Time 8-6
8-7 PVP-I2 Calibration with PVP of Different Chain Lengths 8-8
8-8 PVP Removal by UV/H_2O_2 8-9
8-9 Biodegradability of Untreated Waste and Individual Compounds 8-10
8-10 Long-Term Respirometric Test Results for PVP Solutions 8-11
8-11 Long-Term Respirometric Test Results for Simulated Waste 8-12

A-1 Long-Term Respirometric Results for DEG A-2
A-2 Long-Term Respirometric Results for TMB A-3
A-3 Long-Term Respirometric Results for EB A-3
A-4 Long-Term Respirometric Results for MC A-4
A-5 Long-Term Respirometric Results for VDC A-5
A-6 Results of Inhibition Tests on PCP A-6
A-7 Respirometric Studies for Mixture of Glucose and Glutamic Acid A-7
A-8 Volatilization of Chemicals During Aeration A-8
A-9 Removal of VOCs During Fenton Treatment A-8

LIST OF ACRONYMS

AE	alcohol ethoxylate
AOP	advanced oxidation processes
AP	aminophenol
APEs	alkylphenolethoxylates
AS	Lyophilized activated sludge
BOD	biochemical oxygen demand
CB	chlorobenzene
COD	chemical oxygen demand
CP	chlorophenol
DAP	diaminophenol
DCDE	Dichloro diethyl ether
DCP	dichlorophenol
DEG	diethylene glycol
DI	deionized
DMF	dimethyl dormamide
DNP	dinitrophenol
DO	dissolved oxygen
DOC	dissolved organic carbon
EB	ethyl benzene
EC_{50}	the concentration that kills 50% of the test organisms
EO	ethylene oxide
ETBE	ethyl tertiary butyl ether
F/M	food to microorganism ratio
GC	gas chromatography
GC/MS	gas chromatography/mass spectrometry
Gly	glycerin
H_2O_2	hydrogen peroxide
hr	hour
HRT	hydraulic residence time
IC	inorganic carbon
K_H	Henry's Law constant
k_La	mass transfer coefficient
Kow	octanol/water partition coefficient
K_S	Monod constant, half-saturation constant
L	liter
LAS	linear alkylbenzene sulphonate
LSAEs	linear secondary alcoholethoxylates
M	molar
MBR	membrane bioreactor
MC	methylene chloride
mg	milligram
min	minutes
MLSS	mixed liquor suspended solids
mM	millimolar or millimole

MMP	2-methoxy-2-methyl propionaldehyde
MS	mass spectrometry
MSS	mineral salt solution
MST	1,3,5-trimethyl benzene
MTBE	methyl tertiary butyl ether
N	normal
NMP	n-methylpyrrolidone
NP	nitrophenol
npTOC	non-parent TOC: total TOC - TOC attributable to parent compound
OECD	Organization of Economic Cooperation
$OH^{\bullet}$	hydroxyl radical
O_3	ozone
PCB	polychlorinated biphenyl
PCP	pentachlorophenol
PDC	dichloro propylene
PFBHA-HCl	pentaflurobenzyl)-hydroxylamine hydrochloride
PO	propylene oxide
PPGs	ploypropylene glycols
PPPD	pulsed-power plasma discharge
PTFE	polytetrafluoroethylene
PVP	polyvinylpyrrolidone
q_{max}	Monod constant, maximum respiration rate
RAS	return activated sludge
RCRA	Resource Conservation and Recovery Act
s	seconds
S	substrate concentration
SS	suspended solids
SBR	sequential batch reactor
SDSU	San Diego State University
TBA	tertiary butyl alcohol (*tert*-butyl alcohol)
TBF	tertiary butyl formate (*tert*-butyl formate)
TBOD	theoretical biochemical oxygen demand
TC	total carbon
TCE	trichloroethylene
TEGDME	glycol dimethyl ether
THM	trihalomethane
THMFP	THM formation potential
TMB	1,2,4-trimethylbenzene
TMSD	trimethylsilyl diazomethane
TOC	total organic carbon
TOD	theoretical oxygen demand
UCR	ultimate COD removal
UDR	DOC removed
UV	ultraviolet radiation
VDC	dichloro propylene
VOA	volatile organic analysis
WAO	Wet air oxidation

EXECUTIVE SUMMARY

Biological treatment is a common and usually the least costly process used to treat organic matter in wastewaters. However, certain organic compounds cannot biodegrade within the typical time frame of biotreatment processes and hence pass through the process without removal. Other types of organic compounds might upset the biotreatment process through toxicity or inhibition of the biomass activity. As a result, industrial wastewaters containing such organic compounds can neither be discharged to municipal sewer systems nor treated by onsite biotreatment facilities, thereby often significantly increasing the cost of disposal.

The main objective of this research is to investigate the capabilities of three chemical oxidation processes as pretreatment technologies with the goal of making wastewaters containing persistent organic compounds amenable to biotreatment. The processes investigated are ozonation, ultraviolet radiation/ hydrogen peroxide (UV/H_2O_2) and Fenton reaction. The studies were focused on two organic compounds: dichloro diethyl ether (DCDE) and methyl tertiary butyl ether (MTBE). Synthetic solutions, real groundwater, and industrial wastewater were used for experimentation.

The experimental method involved oxidation of solutions of the target organic compounds at various percentages by the three oxidation processes. Oxidation products were identified and quantified, although closing the full material balance was not always possible. The pre-oxidized solutions of the organic compounds were subjected to biodegradation and toxicity studies. The second major objective of this study was to develop an experimental protocol to effectively test the biodegradability of the oxidized organic compounds by activated sludge. This was accomplished by performing four respirometric tests and one bench scale sequencing batch reactor (SBR) test. Activated sludge was used as the test culture. Tests were conducted by both non-acclimated activated sludge and activated sludge acclimated to the target (parent) compounds. The biodegradability of the oxidized solutions was tested by: 1) short-term tests (about 30 minutes), 2) mid-term tests (up to 24 hours), and 3) long-term tests (up to 14 days). Additional short-term inhibition tests were performed respirometrically to evaluate if the oxidation by-products of the target compounds would exhibit toxicity on activated sludge that use acetate as the major substrate. During the short-term, long-term, and inhibition tests, the biodegradation of organic matter was followed by continuous monitoring of oxygen consumption. The long-term tests also involved the continuous measurement of carbon dioxide. In the mid-term and SBR tests, the concentrations of total organic carbon (TOC), chemical oxygen demand (COD) and mixed-liquor suspended solids (MLSS) were followed as a function of time.

All three chemical oxidation processes were able to transform DCDE and MTBE to a variety of oxidation by-products. Due to considerable amounts of MTBE removal by physical stripping, ozonation was not considered to be an acceptable method for treatment of MTBE. Oxidation by-products of all three oxidation methods were found to be significantly more biodegradable than the non-oxidized parent compounds. It was determined that the results of all the tests used in this study provided valuable information about different aspects of biodegradability and toxicity of oxidation products.

The short-term respirometric test was simple to execute. All the oxidized samples of DCDE and MTBE at every level of oxidation showed an immediate increase in net oxygen consumption, indicating the formation of biodegradable compounds from chemical oxidation. This test proved to be a valuable tool for preliminary screening. However, it will be difficult and most likely misleading to try to extract any other information from this test (e.g., comparison of behavior of acclimated versus non-acclimated cultures, obtaining Monod constants, trying to understand the inhibition or toxicity of the oxidation by-products) due to many limitations of the test.

The long-term test was more tedious. However, it showed that increasing the level of oxidation increased the biodegradability of the oxidation by-products for both DCDE and MTBE. In other words, higher oxidation levels produced by-products that were more biodegradable. Furthermore, all three oxidation processes provided very positive results, except Fenton treatment of DCDE solutions.

The mid-term tests were designed to provide information about the amount of biodegradation of the oxidation by-products within 6 hours—a typical residence time in a treatment plant–in terms of TOC and COD removal. This test showed 25% biodegradation for DCDE oxidation by-products and about 57% biodegradation for MTBE oxidation by-products. This test did not allow the acclimation of the cultures to the oxidation by-products and, as a result, showed the minimum expected biodegradability after chemical oxidation. For example, the SBR results with oxidized DCDE solutions indicated that the by-products of chemical oxidation could be biodegradable by 90%–100% after going through an acclimation period.

The toxicity tests were conducted to evaluate if the chemical oxidation by-products can cause inhibition of the microorganisms that use a readily biodegradable substrate. This test mimicked a municipal activated sludge process that received chemically oxidized industrial wastewater. The results of the test with acetic acid as the primary substrate showed that oxidation by-products of MTBE were not toxic to the biomass at all. Oxidation by-products resulting from UV/H_2O_2 treatment of DCDE did not show any toxicity either. However, oxidation by-products of ozonation and Fenton treatment of DCDE caused considerable inhibition. Long-term tests confirmed the toxicity of the by-products of Fenton-treatment of DCDE, but not of the ozonation. However, the SBR tests showed no toxicity for the DCDE oxidation by-products. This was probably because the relatively large biomass concentrations were used in SBRs (2,000–3,000 mg/L) compared to the mid-term tests (400 mg/L). In addition, the inhibition tests used acetic acid as the main substrate, whereas the SBRs used a more complex feed that is more representative of municipal wastewater.

The SBR test was quite successful in providing most of the needed information in one single test. The SBR studies of the samples of oxidized DCDE showed that almost always the reactors that received the oxidized DCDE produced less removal of COD and TOC compared to the controls. However, the difference in COD and TOC removal efficiencies of these reactors and the controls diminished over time showing considerable acclimation of the organisms to the oxidation products. It was estimated that the reactors initially biodegraded 50%–70 % of the oxidation by-products. However, after acclimation, these percentages increased to 80%–100%. No toxicity or inhibition of oxidation products was apparent.

The SBR studies of samples containing MTBE oxidation by-products showed that the reactors that received the oxidation by-products produced less removal of COD and TOC compared to the controls. Based on the results it was estimated that only about 50% of the oxidation products biodegraded under the experimental conditions. This result agrees with the mid-term respirometric results that about 57% of the by-products were biodegradable. There was no further acclimation over a 12-day period.

Tests conducted on MTBE-containing groundwater and wastewater that contained nonbiodegradable polyvinylpyrrolidone (PVP) showed significantly improved biodegradation after chemical oxidation based on long-term respirometric studies. The SBR tests showed that about 63% of the oxidation by-products of groundwater were biodegradable. After oxidation, the PVP solutions showed highly increased biodegradability as measured by carbon dioxide formation. Similarly, simulated waste containing oxidized PVP also exhibited improved biodegradation even after 45% oxidation.

Overall, the SBR test was quite successful in providing most of the needed information in one single test. The information obtained through SBR included:

- Whether the chemical oxidation by-products were biodegradable under the operational conditions of a typical activated sludge plant;
- The percentage of the oxidation by-products that was biodegradable;
- Whether the chemical oxidation by-products caused any inhibition on the biomass activity; and
- Whether the biomass showed any acclimation during the treatment to the oxidation by-products.

This project clearly showed that integration of chemical oxidation processes with biotreatment might lead to effective handling of "problem" wastewaters. Furthermore, the experimental protocol developed and used in this study might prove to be a valuable tool for comprehensive evaluation and assessment of the integration of chemical oxidation with biotreatment processes.

CHAPTER 1.0

INTRODUCTION

1.1 Rationale for Research

The biodegradability of many naturally occurring (biogenic) and synthetic (xenobiotic) organic chemicals has been studied extensively in order to understand the fate of these chemicals in the natural environment as well as in engineered systems, such as biological treatment plants. Although most biogenic compounds are fairly degradable, a considerable number of xenobiotic compounds were found to be *persistent*. A compound might be persistent due to being *recalcitrant*, or not being able to go through biodegradation under a given set of conditions; i.e., these compounds may inherently be biodegradable under the right conditions (Grady, 1985). In an engineered system, such as an activated sludge process, xenobiotic compounds may exert three major effects (Volskay and Grady, 1990):

- First, they may be inhibitory to their own biodegradation through substrate inhibition.
- Second, xenobiotic compounds can influence the rate at which at which biogenic organic matter is used by microorganisms capable of degrading both biogenic and xenobiotic substrates.
- Third, xenobiotic compounds can inhibit the biodegradation of the biogenic organic matter by microorganisms that are incapable of degrading the xenobiotic compounds.

As a result, xenobiotic compounds may pass through the biological process without any transformation, and, furthermore, they may cause process to fail. Certain polymeric compounds, e.g., polyethylene glycol, may be persistent because of being non-bioavailable. Furthermore, most hydrophobic xenobiotic compounds easily partition to the biosolids and leave the biological treatment system as part of the sludge, creating potential problems in sludge treatment and disposal options. Finally, relatively volatile xenobiotic compounds tend to strip during biological treatment, especially in surface-aerated activated sludge reactors, resulting in a potential air-borne health hazard.

The discharge of the majority of the persistent, hydrophobic, and volatile xenobiotic compounds to municipal or industrial systems that use biological systems is often restricted or prohibited. For example, the 1996 Toxic Release Inventory for California indicates that the chemical industry, the largest contributor to the toxic waste discharge in the State, could discharge only about 47% of the waste to the wastewater treatment plants (WWTP). Similarly, the petroleum industry, the second largest contributor, discharges only about 25% to the WWTPs, while treating about 10% on-site by biological processes. The remaining large percentage of the chemicals has to be taken off-site and treated or disposed of by costly and potentially liable alternatives such as underground injection and land disposal.

Therefore, there is need for technologies that can be used at the source to treat persistent xenobiotic compounds in order to modify their chemical characteristics to render them treatable in a biological system without any adverse effects. Physical processes by nature cannot alter the chemical structure of the compounds. Certain chemical processes, such as hydrolysis, dehalogenation by chemical reduction, e.g., zero-valent iron, or alkaline polyethylene glycol (APEG) processes have limited applications (Gurol, 1993). For example, APEG can treat only polychlorinated biphenyls (PCBs) and other halogenated aromatic compounds in oil and soil, whereas, most reduction processes require a strong reducing environment, have very slow reaction rates, and could work only for highly oxidized chemicals, e.g., certain azo dyes. Chemical oxidation processes, on the other hand, have demonstrated capability for destruction of a variety of hazardous compounds. The original compounds are converted through oxidation to smaller, generally biogenic, and as a result biodegradable intermediate compounds, e.g., carboxylic acids, alcohols, and aldehydes. The research focuses on the use of chemical oxidation technologies for pretreatment of various organic contaminants to improve their removal in subsequent biotreatment units.

1. 2 Background

1. 2.1 Chemical Oxidation Processes

1.2.1.1 General Review

Organic compounds can be oxidized at ambient temperatures and atmospheric pressure by a variety of oxidants, including ozone, hydrogen peroxide, and potassium permanganate, by irradiation using ultraviolet (UV) light, gamma rays from radioisotopes, e.g., Co-60, and by high energy electrons from electron accelerators. These non-thermal chemical oxidation technologies are generally applicable for relatively dilute aqueous waste (< 0.1% organic concentration), which constitutes about 90% by volume of the hazardous waste generated in the United States by industrial activities (Baker and Warren, 1992).

Ozone is one of the more commonly used oxidants, yet it is selective in its reactions with organic compounds. It reacts with certain organic chemicals at a very fast rate; for example, the oxidation rate constants for hydroquinone and phenol, are respectively 1.1×10^6 and1.3×10^3 M^{-1} s^{-1} (Gurol and Nekouinaini, 1984). On the other hand, the reaction of ozone with certain chemicals is very slow; for example, chlorobenzene and perchloroethylene react with ozone with the rate constants of 0.75 and 0.1 M^{-1} s^{-1}, respectively (Hoigne and Bader, 1983). Hence, ozonation may be an excellent pretreatment technique for certain organic chemicals, yet it may not work at all for others.

The so-called advanced oidation processes (AOPs) have been developed to overcome the limitation of ozonation by relying on the generation of hydroxyl radical (•OH), which is a very strong and non-selective oxidant for organic compounds. The rate constant for the oxidation reactions with •OH, is generally in the range of 10^7 to 10^{10} $M^{-1}s^{-1}$, which leads to instantaneous reactions. Consequently, AOPs are capable of oxidation of almost any organic chemical at a fast rate that is determined primarily by the rate of •OH formation in the system, and the amount and type of the other chemicals that compete with the target organic compound for •OH. Hence, it is technically possible to achieve complete degradation of the organic matter, i.e., *mineralization*, by using AOPs.

Among all the AOPs, the use of UV radiation in conjunction with ozone or hydrogen peroxide was observed to be the most commonly used technology. Currently, well over a hundred pilot and full-scale plants are using UV/O_3 and UV/H_2O_2 processes successfully for removal of organic substances from surface and ground waters. These processes primarily use low or medium pressure mercury arc lamps, although xenon and excimer lamps have also been very successful in laboratory experimentations. In addition, several catalytic processes have demonstrated success in generating •OH, including the UV/TiO_2, Fenton reaction and H_2O_2 with iron oxide minerals. Fenton's reaction (or reagent) involves the use of H_2O_2 with ferrous salts. Recent studies indicate that similar oxidative effect can be obtained by using ferric salts, which only delay the oxidation process by a few minutes (Lin and Gurol, 1994). Hydrogen peroxide can also interact with iron oxide minerals that can be in granular or colloidal form to generate •OH (Lin and Gurol, 1998). Two novel techniques, the ultrasound and pulsed-power plasma discharge (PPPD), also produced promising results in laboratory research (Hoffman, 1993).

In addition to the non-thermal technologies mentioned above, two processes that rely on high-temperature and pressure, namely, wet air oxidation (WAO) and supercritical fluid oxidation have also been very successful in destruction of many hazardous compounds. These processes are generally applicable for treatment of wastes that have high concentrations of organic matter (1.5%–10%). The WAO process requires an operational temperature range of 175–320°C and pressure range of 20–200 atm (Copa and Gitchel, 1988). Although 80–90% of the organic matter (as chemical oxygen demand, COD) is easily oxidized, many chlorinated and nitro substituted compounds may show resistance, especially without additional catalysts, e.g. Cu (Dietrich et al., 1985). Most of the remaining organic matter is in the form of carboxylic acids that are readily biodegradable (Hao et al., 1994; Baillod et al., 1985). Supercritical fluid oxidation on the other hand, requires extreme conditions for proper operation, i.e., temperatures of 450–650°C and pressures of 250–300 atm. The process is exceptionally efficient in destroying the organic matter in seconds, leaving very little organic residue that needs further treatment.

Table 1-1 provides a comparison of these oxidation technologies for application in aqueous phase (Gurol, 1993). Based on the known capabilities, cost considerations and the extent of development, the considerations that are summarized in Table 2-1, it was decided to select the following three processes as pretreatment methods in this study: 1) ozonation, 2) UV/H_2O_2, and 3) Fenton reaction.

The major reasons for selection of the three oxidation methods are summarized below:

- All three processes have demonstrated capabilities for destruction of various hazardous compounds.
- All three processes have full-scale applications, indicating their technical and economic feasibility.
- All three processes can convert refractory and inhibitory organic compounds to biogenic and therefore biodegradable compounds with low probability of formation of toxic intermediate products.
- The reaction mechanisms of these processes are fairly well established in terms of the primary oxidants involved. For example, while ozone is the primary oxidant in the ozonation process, •OH is known to be primarily responsible for oxidation in the

UV/H_2O_2 and Fenton reaction (Gurol and Akata, 1996; Gurol and Liao, 1991; Gurol, 1985; Yurteri and Gurol, 1987).

1.2.1.2 Ozonation

Ozone is commonly used for disinfection of water and for removal of taste- and odor-causing chemicals from drinking water sources. It is also a very effective oxidant for certain organic compounds that have double bonds or aromatic rings, such as chloroethylene and phenols. However, many organic compounds resist the attack by molecular ozone. The ozone molecule can react with hydroxide ion and many other chemicals present in natural waters (e.g., iron, humic material) to decompose to hydroxyl radical (•OH). This radical, being a strong and indiscriminate oxidant, can oxidize organic compounds efficiently. However, •OH formation rate from ozone can be accelerated by exposing ozone to UV light or with addition of hydrogen peroxide to ozone.

There is no information in the literature on ozonation of dichlorodiethyl ether (DCDE). Information on methyl tertiary butyl alcohol (MTBE) ozonation is also limited. Leitner et al. (1994) studied the oxidation of MTBE and ethyl tertiary butyl ether (ETBE) by ozone and by ozone/hydrogen peroxide (known as peroxone) and reported that ETBE reacted faster than MTBE. Furthermore, peroxone was more effective than ozone alone in removing both compounds. Liang et al. (2001) also reported that peroxone was more effective than ozone alone in oxidizing MTBE to the by-products, including tertiary butyl formate (TBF), tertiary butyl alcohol (TBA), acetone, and aldehydes. Safarzadeh-Amiri (2001) who studied MTBE oxidation by peroxone concluded that MTBE reacted primarily by •OH. It was also concluded that peroxone is a more efficient and less costly process than UV/H_2O_2.

Table 1-1. Comparison of Chemical Oxidation Processes (Gurol, 1993).

Considerations	O_3	O_3/H_2O_2	UV/O_3 UV/H_2O_2	Electron Beam	TiO_2/UV	Ultrasound PPPD	Fenton Reagent	H_2O_2/ Iron Oxide	Wet Air Oxid.	Supercritical Fluid Oxid.
Operational conditions	Ambient	Ambient	Ambient	Ambient	Ambient	Ambient	Ambient	Ambient	175–320°C 20–200 atm	450–650°C 250–300 atm
Broad compound applicability	No	Yes	Yes	Yes	No	Yes	Yes	No	No	Yes
Rapid oxidation of chlorinated organics	No	Yes	Yes	Yes	Yes	Yes	Yes	Yes	No	Yes
Capable of complete mineralization	No	Yes	Yes	Yes	No	Yes	Yes	No	No	Yes
Residuals that need further processing	No	No	No	No	Yes/No	No	Yes	No	Yes	No
Probability of undesirable product formation	Low	Low	Low	Low	Low	High	Low	Low	Low	Low
Easily portable	No	No	Yes (UV/H_2O_2) No (UV/O_3)	No	Yes	Yes	Yes	Yes	No	No
Full scale demonstration	Yes	Yes	Yes	No	No	No	Yes	No	Yes	No
Relative cost	Medium	Medium	Medium	High	Medium	High	Low	Low	Medium	High
Potential use as a pretreatment process	Yes	Yes	Yes	No	No	No	Yes	No	Yes	No

1.2.1.3 Fenton Reaction

Mixtures of iron salts with H_2O_2, which are known as Fenton processes, are potential alternative oxidation methods for organic compounds. Although an electron transfer mechanism for oxidation of organic compounds was proposed in recent literature for these reactions (Sawyer et al., 1996; Bossman et al., 1998), it is generally believed that hydroxyl free radical that is generated through the interactions of H_2O_2 with iron salts, according to the following reactions, is responsible of oxidation of organic compounds (Walling and Clearly, 1977; Walling 1975; Barb et al., 1951).

$$Fe(II) + H_2O_2 \rightarrow Fe(III) + OH^- + \bullet OH \quad (1)$$

$$Fe(III) + H_2O_2 \rightarrow Fe(II) + H^+ + HO_2\bullet \quad (2)$$

$$Fe(III) + HO_2\bullet \rightarrow Fe(II) + H^+ + O_2 \quad (3)$$

$$\bullet OH + H_2O_2 \rightarrow HO_2\bullet + H_2O \quad (4)$$

$$\bullet OH + Fe(II) \rightarrow Fe(III) + OH^- \quad (5)$$

$$\bullet OH + \text{organic compound(s)} \rightarrow \text{product(s)} \quad (6)$$

The •OH reacts with the organic compound(s) of interest at a very fast rate, and as such is believed to be primarily responsible for oxidation of organic compounds in these systems. No information is available in the literature on the rate constant of the reaction between DCDE and •OH. However, MTBE reacts with •OH at a rate constant of $1.6 \times 10^9\ M^{-1}s^{-1}$ as determined by photolysis reactions (Buxton et al., 1988).

Reaction 1 is much faster with a rate constant of about 76 $M^{-1}s^{-1}$ than Reaction 2 that has a rate constant of 0.01–0.02 $M^{-1}s^{-1}$ (Walling, 1975). As a result, ferrous initiated reaction, which is referred to as Fenton reaction, is much faster than the ferric-initiated reaction. However, as soon as the initial ferrous is consumed and transferred into ferric iron, the chain reactions are propagated by the subsequent reactions of ferric, and the rate is controlled by Reaction 2 that regenerates ferrous. To date, as many practical applications have involved the use of H_2O_2-ferric as the H_2O_2-ferrous mixtures. It was reported that although the initial rate of mineralization of organic compounds was higher for ferrous-initiated mixtures, the extent of overall mineralization was the same for both ferrous and ferric initiated reactions (Lunar et al., 2000; Gallard and De Laat, 2001). We conducted our experiments with a ferric salt, specifically $Fe_2(SO)_4$, which is a readily available water-conditioning chemical. Additional studies conducted with ferric chloride were reported here for comparison as well.

Recently, Burbano et al. (2002) reported a study on degradation of MTBE and its oxidation by-products by Fenton's reagent. The major intermediate by-products generated during the reaction were TBF, TBA, methyl acetate, and acetone. It was proposed that TBA was generated by reactions of TBF, methyl acetate was formed directly from MTBE, and acetone was produced by at least three independent pathways.

1.2.1.4 UV/H_2O_2

Hydrogen peroxide when exposed to UV light is known to split into •OH. While hydrogen peroxide is photo reactive over the 185–400 nm wavelength range, short-wave ultraviolet energy (200–280 nm) results in the highest •OH yields. The molar absorptivity of hydrogen peroxide at 253.7 nm is 19.6 M^{-1} cm^{-1} and the quantum yield for •OH production from H_2O_2 at 253.7, 308, and 351 nm is approximately 1.0 mole of peroxide decomposed per mole of light photons absorbed (Baxendale and Wilson, 1956).

The reaction mechanisms depicted below describes the photolysis of H_2O_2 in pure water.

$$H_2O_2 + \text{UV light} \text{ ------}\rightarrow 2\ \bullet OH$$

$$\bullet OH + H_2O_2/HO_2^- \text{ ------}\rightarrow HO_2\bullet + H_2O$$

$$HO_2\bullet + H_2O_2 \text{ ------}\rightarrow \bullet OH + H_2O + O_2$$

$$HO_2\bullet + HO_2\bullet \text{ ------}\rightarrow H_2O_2 + O_2$$

Again, •OH generated in the reaction is believed to be the primary oxidant for the organic compounds.

Oxidation of DCDE by this process was reported earlier by a group of scientists (Huang and Liu, 1999; Li et al., 1995). The process was applied for treatment of an aquifer that was contaminated with a landfill leachate. DCDE was one of the 17 major chlorinated hydrocarbons found in the groundwater with concentrations of up to 1200 ppm. It was proposed to remove the volatile compounds from the groundwater by air stripping, and the low-volatility compounds such as DCDE by UV/H_2O_2 oxidation. It was reported that DCDE at a concentration of 200 ppm was oxidized effectively in the presence of H_2O_2 at an initial molar ratio of 1/10 with a medium pressure mercury lamp. The lamp radiated 175 W of energy, 40–48% of which was in the UV range.

Oxidation of MTBE by UV/H_2O_2 was studied by Wagler and Malley (1994), who reported 95% removal of MTBE within 40 minutes. Chang and Young (2000) conducted the tests in a recirculating batch reactor in the presence and absence of UV radiation to separate the effects of MTBE loss to system components and photolytic processes. Experiments were conducted at initial peroxide/MTBE molar ratios of 4:1, 7:1 and 15:1. The UV/H_2O_2 treatment resulted in >99.9% removal of MTBE with the major purgeable by-product identified as TBF. The second order rate constant for the degradation of MTBE from the hydroxyl radical was estimated to be 3.9×10^9 $M^{-1}s^{-1}$. The second order rate constant for the reaction of TBF with the hydroxyl radical was estimated to be 1.2×10^9 $M^{-1}s^{-1}$. The yield for the formation of TBF from the reaction of MTBE and the hydroxyl radical was 0.28. After 1 hour of treatment the TBF remaining exceeded the remaining MTBE by as much as a factor of 35. Stefan et al. (2000) also studied the degradation of MTBE by UV/H_2O_2 process, and reported TBF, TBA, 2-methoxy-2-methyl propionaldehyde (MMP), formaldehyde, acetone, and methyl acetate as the primary by-products.

1.2.2 Use of Chemical Oxidation for Pretreatment Prior to Biotreatment

1.2.2.1 Literature Review

Scott and Ollis (1995) provided an excellent literature review on the application of pre-chemical oxidation to enhance the biodegradability of various chemicals and waste types. The review indicates that compounds that are recalcitrant mainly because of their large polymeric structures, e.g., polyethylene glycol (Suzuki et al., 1976) would benefit the most from preoxidation that breaks these compounds into smaller and more biodegradable fragments. Various other recalcitrant compounds, including pesticides (Kearney et al., 1988; Somich et al., 1990), herbicides (Somich et al., 1988), and munitions (Kearney et al., 1983) showed significant enhancement in biological activity upon chemical oxidation. In addition, numerous biodegradable compounds that possess some degree of toxicity or activity inhibition, e.g., methyl vinyl ketone, dichlorophenol (Manilal et al., 1992), and substituted aromatics (Gilbert, 1987) degraded into less toxic and more biodegradable intermediates. Other studies have indicated significant improvement in biodegradation rate of pentachlorophenol (PCP) and trichloroethylene (TCE) upon treatment with hydrogen peroxide and Fenton's Reagent (Gurol and Ravikumar, 1992; Ravikumar and Gurol, 1991). Both PCP and TCE were partially oxidized in soil environment before being subjected in continuous-flow column studies to biodegradation by activated sludge cultures. The significant enhancement observed in the rate of biodegradation, which was monitored by the disappearance of the compounds and the accumulation of chloride ion, was attributed to co-metabolic activity (Ravikumar, 1992). In experiments performed by Kiwi et al. (1993), photo-Fenton oxidation was used to determine the biodegradability of oxidized 2-nitrophenol. After 100% oxidation 2-nitrophenol was found not as biodegradable as the parent compound. The BOD_5/COD (biochemical oxygen demand$_5$/chemical oxygen demand) values for non-oxidized and oxidized solutions were, respectively, 0.61 and 0.25.

Since the review by Scott and Ollis (1995), a number of studies have been published using combined chemical and biological processes. Recently, Marco et al. (1997) used 2,4-dicholorophenol as a model non-biodegradable compound. Acetic and formic acids were also added to the simulated wastewater. The results showed that biodegradable compounds such as formic acid, which is a common by-product of oxidation of most organic compounds, could slow the oxidation process of the targeted organic compound by competing for ozone. Bolduc and Anderson (1997) conducted experiments on three aromatic compounds, *m*-dinitrobenzene, diphenylamine, and resorcinol, by using combined photocatalytic pre-oxidation and biodegradation processes. The photocatalytic treatment was performed in a TiO_2 slurry reactor containing one of the chemicals in aqueous solution. Biodegradability was evaluated by estimating the ratio of 1-day BOD to total organic carbon (TOC). For toxicity, the Microtox test, which utilizes bioluminescent *Photobacterium phosphoreu,* was used. The intermediate compounds formed during photocatalytic oxidation showed higher inhibition than the parent compounds for *m*-dinitrobenzene and diphenylamine, and less inhibition for resorcinol. For solutions of *m*-dinitrobenzene and diphenylamine, improvement in the BOD_1/TOC ratio was achieved, but it required the chemical mineralization of at least 80% of the organic carbon originally present in the water.

In a study performed by Adams et al. (1997), batch tests were conducted to analyze the effects of ozonation on the biodegradation of 4-chlorophenol (4CP), 2,4-dichlorophenol (DCP), 4-nitrophenol (4NP), 2,4-dinitrophenol (DNP), 4-aminophenol (4AP), and 2,4-diaminophenol (DAP). Each batch assay was operated for about 300 hours and initiated with a microbial seed, the target compound (99% oxidized or unoxidized), and a nutrient solution. The ultimate COD removal (UCR) was analyzed in order to determine the biodegradation of these compounds and their by-products. Initially, DNP, 4NP, DCP, and 4CP (nitro- and chloro-phenols) were found to be recalcitrant at a concentration of 200 mg/L as COD but readily biodegraded at a concentration of 30 mg/L as COD (80%–97%), which shows that these chemicals were in fact toxic at high concentrations, yet biodegradable. After ozonation of the 200 mg/L as COD solutions, biodegradation increased from none to 72%, 52%, 51%, and 44% for DNP, 4NP, DCP and 4CP, respectively. On the other hand, aminophenols (4AP, DAP) were highly biodegradable (95%+) at both 30 and 200 mg/L as COD concentrations before oxidation. After oxidation the biodegradation of 4AP and DAP reduced to 57% and 71%, respectively. Microtox tests were also performed in this study and showed a decrease in toxicity after oxidation of each original compound. The largest change in effective concentration (EC_{50}, the concentration that killed 50% of the organisms) for compounds that increased in biodegradation after oxidation was with 4-chlorophenol, which ranged from less than 1 mg/L as COD before oxidation to greater than 35 mg/L as COD after oxidation. Although 4AP and DAP showed a reduction in biodegradation after oxidation, the EC_{50} increased from 30 mg/L as COD to approximately 300 mg/L as COD for both chemicals.

Beschkov et al. (1997) investigated the degradation of triethylene glycol dimethyl ether (TEGDME) by ozonation coupled with UV light or H_2O_2 in order to determine if the by-products would be more biodegradable. In the presence of UV light, ozonation increased the BOD_5 of a 100 mg/L TEGDME solution from 0 to 20 mg/L. Without UV light, ozonation only produced a BOD_5 of about 4 mg/L. Djongalic and Levec (1998) compared biodegradability of diluted aqueous azo dye solutions after wet oxidation with and without a compounded catalyst containing 42% by weight of copper oxide, 47% zinc oxide and 10% aluminium oxide. They found that intermediate products were almost identical with and without the catalyst but that the catalytic oxidation yielded 50% lower concentration of the by-products and resulted in higher biodegradability. Ito et al. (1998) compared the efficiency of ozonation with AOPs, including O_3/UV, O_3/H_2O_2, H_2O_2/UV and O_3/H_2O_2/UV in systems combined with biodegradation for the removal of trihalomethanes (THMs) precursors in treated sewage effluents. They estimated biodegradability by the differences in dissolved organic carbon (DOC), UV260, and THM formation potential (THMFP) before and after incubation for 7 days at 20°C. It was found that the contribution of ozonation alone to the mineralization of DOC was small, but ozonation improved the subsequent biodegradability of residual DOC. On the other hand AOPs were more effective for both mineralization of organic compounds and the increase in the biodegradability. Also, if longer oxidation times were allowed, the extent of mineralization increased but not the biodegradability, probably due to the decomposition of biodegradable compounds.

Kornmüller and Wiesmann (1999) used continuous ozonation of the polycyclic aromatic hydrocarbons benzo(e)pyrene and benzo(k)fluoranthene in oil/water emulsions. They measured biodegradation by analyzing concentrations of the targeted chemicals and the ozonation products and found that the ozonation by-products were amenable to biodegradation. Kitis et al. (1999) studied the effect of Fenton's reagent pretreatment on the biodegradability of three non-ionic surfactants and performed an economic analysis of each treatment. They quantified biodegradability by plotting the dissolved organic carbon (DOC) versus time and obtained the ultimate percent DOC removed (UDR) and the time to achieve maximum UDR. Fenton's reagent was found highly effective in enhancing the biodegradability of these chemicals. Kitis et al. (2000) used ozone/hydrogen peroxide treatment for pre-oxidation of ethylene oxide/propylene oxide (EO/PO)block copolymers, ploypropylene glycols (PPGs), linear secondary alcoholethoxylates (LSAEs), and alkylphenolethoxylates (APEs). Oxidized samples were bioassayed in aerobic batch systems. Increasing oxidant dosages consistently increased both the rate and the extent of biodegradation.Yeber et al. (1999) compared reactions with suspended and supported catalysts, TiO_2 and ZnO, in photocatalysis to degrade two bleaching effluents from wood processing in order to reduce toxicity and to improve biodegradability. They claimed that after one minute of photocatalysis the effluent improved the growth of microorganisms in comparison to the untreated effluent. Also, they claimed acute toxicity reduction of 50%. Adams and Kuzhikannil (2000) studied the effect of UV/H_2O_2 preoxidation on the aerobic biodegradability of quaternary amine surfactants. They quantified improvement in biodegradation by measuring ultimate dissolved organic carbon removal. Preoxidation effectively enhanced the biodegradability of alkyldimethylbenzyl ammonium chlorides but had little effect on the biodegradability of dioctyl-dimethyl ammonium chloride. They suggested that significant differences in biodegradability after oxidation are related directly to the structure of the oxidation by-products. Beltran et al. (1999) performed laboratory-scale experiments to investigate the impact of ozonation on the degradability of wine distillery wastewaters. Changes in the BOD/COD ratio were followed to determine the ozonation conditions that improved the biodegradability of the wastewater tested. They found that ozone doses higher than optimum were not capable of converting refractory organics to more biodegradable compounds. Chamarro et al. (2001) studied the biodegradation of acetic acid, formic acid, phenol, 4-chlorophenol, 2,4-dichlorophenol, and nitrobenzene in aqueous solution, with the focus on the dependance of the amount of oxidant used. They used Fenton's reagent as the oxidation method and BOD_5/COD and BOD_{21}/COD ratios as indicators of improvement in biodegradability. They showed that the total elimination of organic carbon by pre-oxidation would require large amounts of oxidant and long residence times.

1.2.2.2 Deficiencies in Prior Work

More than 80 publications reviewed by Scott and Ollis (1995) and several articles published more recently revealed that the use of following oxidation processes caused significant enhancement in the biodegradability of the individual compounds or the industrial wastewaters used in these studies: O_3, UV/O_3, O_3/H_2O_2, UV/ H_2O_2, Fenton's reagent, TiO_2/UV or sunlight, WAO, gamma radiation, electron beam, and various catalytic oxidations. These studies, however, have provided very little fundamental information regarding the reaction mechanism, kinetics, water quality conditions, competing reactions, etc. For example, reactions by molecular O_3 were not differentiated from the reactions of •OH, and the quantity of the oxidant used in the reactions in terms of O_3 or •OH has not been reported. Very few of these studies used a quantitative method to describe the amount of oxidant used versus the extent of the organic compound oxidized. Consequently, it is very difficult to relate the data collected for each oxidation method for comparison of the capabilities of the processes. As a result, the information provided in these studies cannot be used in any predictive way for evaluation of the process capabilities under different experimental or operational conditions, or for other types of organic contaminants or waste types.

The review of the existing literature does not clarify why only about 5–10% oxidation might be sufficient to observe significant enhancement in the biodegradation of certain organic compounds, and yet, for others, all of the original compound would need to be transformed to intermediates to be able to observe any enhancement at all. Comparison of the existing data is further complicated because some of the investigators used acclimated and others nonacclimated cultures, some used pure cultures, and still others used anaerobic conditions.

In very few of these studies an attempt was made to identify the intermediate compounds, although the intermediates were not quantified. Even the techniques used in determining the extent of biodegradability varied from study to study. Some used gross estimators, e.g., BOD, COD, and TOC; some used substrate destruction or oxygen uptake. Some investigators used batch reactors, and others designed continuous-flow activated sludge systems or chemostats.

Only a few of the studies provided information on the kinetic constants (Lee and Carberry, 1992; Bowers et al., 1989), while the majority reported qualitative and therefore subjective information on the biodegradability of the test compounds. There has been no effort to differentiate between the biodegradability of the parent compound versus the intermediates of chemical oxidation. Furthermore, it is not clear whether the improvement in biodegradability has been due to the reduction in the concentrations of the inhibitory compounds by chemical oxidation or to the production of biodegradable intermediate compounds.

In addition, the majority of the existing work relied for assessment of biodegradability on the BOD test in which a small fraction of the chemically oxidized solution was diluted by several times and then subjected to a 1- to 5-day test during which oxygen is consumed by the seed microorganisms. This is obviously a crude method that cannot determine whether the parent compound or the oxidation intermediates could be toxic to the microorganisms. This method cannot provide rate information about the biodegradation of the parent compound and intermediate products. The need to develop a methodology that can provide reliable information is clear.

1.2.3. Test Compounds

Thirteen organic compounds listed in Table A-1 (Appendix A) were proposed originally as the test compounds. However, it was later decided at a meeting with the project subcommittee members to change the scope of the study to focus on two chemicals in detail: DCDE and MTBE. The purpose of this change was to emphasize the understanding of the fundamentals of the improvement of biodegradability upon oxidation and the development of a procedure, which can be tested on a smaller number of chemical compounds.

DCDE is widely used in the United States in the manufacture of pesticides and pharmaceuticals as a solvent and cleaning fluid, as a constituent of paints and varnishes, and in the purification of oils and gasoline. Its production in 1986 was reported to be 1.2 tons (ATSDR, 1989). DCDE is nonbiodegradable in conventional biological treatment systems (Lewis, 1991). It is reactive, flammable, a mutagen, and a probable carcinogen at a dose level of 0.3 μg/day (Fisher, 2000). Its inhalation can cause central nervous system depression (ATSDR, 1989). Because of these properties, DCDE is present on the "Extremely Hazardous Substances" list and the "Community Right to Know" list. In addition, DCDE is classified as a hazardous chemical under the Resource Conservation and Recovery Act (RCRA), which mandates that wastes containing DCDE above certain levels be disposed of in an authorized RCRA facility.

MTBE has been used as a gasoline oxygenate in the United States for over 20 years. It replaced lead in gasoline and is used to reduce air pollution and increase octane ratings. MTBE compromises as much as 15% by volume of gasoline, and in 1997, it was the second highest chemical by volume produced in the United States (Mitani et al., 2002). MTBE is a suspected carcinogen and has been shown to complicate reproduction (Fisher, 2000). An LD_{50} by oral ingestion and an LC_{50} by inhalation for a mouse is 5960 μL/kg and 141 $g/m^3/15$ min, respectively (Fisher, 2000). MTBE has been observed to be biodegradable by numerous species of microorganisms, including microorganisms present in a typical activated sludge, but the volatility of MTBE is believed to control the fate of this chemical (Mo et al., 1997). Microorganisms in a conventional aerobic activated sludge process may not have sufficient time to biodegrade the chemical.

Additional known properties of DCDE and MTBE are listed in Table 1-2.

Table 1-2. Properties of the Test Compounds.

Compounds	Formula	M.W.	Physical Properties[1] Solubility in water mg/L	Physical Properties[1] *$logK_{ow}$*	Biodegradability[2] TOD g/g	Biodegradability[2] BOD_5 g/g	Biodegradability[2] BOD_5/TOD %	Oxidation Rates Ozone[3] L/mole.sec	Oxidation Rates Hydroxyl[4] radical L/mole.sec
DCDE	$C_4H_8OCl_2$	143.02	10,200	1.29	1.12	0	0	<1.1	-
MTBE	$C_5H_{12}O$	88.2	48,000	1.3	2.73	0.55	20	-	1.6×10^9

DCDE: dichloro diethyl ether; MTBE: methyl tertiary butyl ether.
TOD: theoretical oxygen demand
$logK_{ow}$: log octanol/water partition coefficient
$logK_H$: log Henry's Law constant
(1) Schwarzenbach et al., 1993
(2) Pitter and Chudoba, 1990
(3) Hoigne and Bader, 1981
(4) Buxton et al., 1988

CHAPTER 2.0

OBJECTIVES AND EXPERIMENTAL METHODS

2.1 Objectives

The main objective of this research is to investigate the capabilities of three different chemical oxidation methods (ozonation; UV/H_2O_2; and Fenton reaction) to increase the biodegradability and reduce the toxicity of two test compounds, DCDE and MTBE.

The specific objectives of the project are to:

- Determine the oxidation rate of each test compound by each oxidation process.
- Determine the oxidant and energy requirement for each process to achieve the desired level of oxidation in the test compounds.
- Identify and quantify the oxidation by-products of the test compounds.
- Elucidate the reaction mechanisms based on observed reaction rates, oxidant use, and formation of intermediate products.
- Assess whether the compounds are recalcitrant and/or inhibitory to non-acclimated and acclimated activated sludge cultures prior to chemical oxidation.
- Determine the extent of improvement in the biodegradability of the test compounds oxidized by three oxidation processes at various oxidation levels.
- Compare the biodegradability results obtained by short-term, mid-term, and long-term respirometry tests with toxicity and sequential batch reactor (SBR) results to understand the mechanism for improvement in the biodegradability of the compounds upon oxidation.
- Develop a methodology to be used for assessment of the effectiveness of a pretreatment technique on enhancing the biodegradability of wastewaters containing refractory compounds.

2.2 Experimental Methods

2.2.1 General Approach

The solutions of DCDE and MTBE were prepared at various concentrations in distilled water buffered with bicarbonate/carbonate salts at neutral pH. The solutions were subjected to ozonation, UV/ H_2O_2, or Fenton treatment under various conditions that are presented later in this report. Chemical oxidation of the test compounds was achieved at various percentages of 0% (nonoxidized), 25%, 50%, 75%, and 100%. These were the target oxidation levels, and the exact percentages were determined after the oxidation tests by measuring the remaining concentration of the test compounds. For some experiments, oxidation was continued beyond complete elimination of the test compound. Samples were analyzed for the parent compound, oxidation by-products, chloride ion for DCDE, ozone concentration in liquid and gas phases during ozonation, hydrogen peroxide concentration during UV/H_2O_2, and Fenton treatment, and dissolved iron concentration during Fenton treatment. Additional measurements, e.g., pH and sludge formation, were conducted as detailed in later sections.

Activated sludge cultures were grown and mainained in our laboratory in two separate containers: one for non-acclimated culture, and the second for culture acclimated to the test compounds. These cultures were used to run the respirometric tests for assessment of biodegradation and toxicity. In addition, SBRs were used to grow activated sludge on non-oxidized and oxidized solutions of the test compounds.

The biodegradation studies were conducted by using three different tests. The first test, short-term biodegradation, involved the measurement of biomass respiration rate by oxygen probes over a period of 30 minutes. The mid-term biodegradation tests were conducted over a period of up to 24 hours. The mid-term experiments were designed to present data in terms of degradation of the test compound and the biodegradable organic matter (TOC and COD measurements). The third method, long-term respirometric tests, was conducted over a period of two weeks to obtain information on both biodegradability and toxicity of the by-products of oxidation and of the parent compounds. The respirometric data in terms of oxygen consumption and carbon dioxide production over the two-week period allowed us to see if prolonged exposure would result in microorganisms developing the necessary enzymes for degradation of the parent compounds and the oxidation by-products or would result in higher toxicity.

In this report the words toxicity and inhibition were used frequently and interchangeably. However, inhibition usually refers to the degree to which the chemical of interest interferes with microorganisms' ability to break down another readily biodegradable compound present. Toxicity, on the other hand, addresses how the chemical of interest destroys cultures without the other readily biodegradable chemical present. Because plenty of naturally occurring, organic compounds exist in activated sludge systems, it is practical to check inhibition (OECD, 1987). However, in receiving waters, where food is scarcer, the threat of toxicity is more relevant (US EPA, 1993). In assessing inhibition on activated sludge, short-term respirometric tests were performed in the presence of readily biodegradable acetate. A modified version of OECD Method 209, "Activated Sludge, Respiration Inhibition Test" was applied. The respiration rate of the system was compared to one with only activated sludge and acetate to see if DCDE caused a reduced respiration rate.

In addition to the respirometric tests described above, bench scale simulated SBRs were used with activated sludge to obtain better insight to the behavior of a compound in a conventional biological system after being subjected to pre-oxidation. Oxidized chemical solutions and a synthetic sewage solution (Syntho) containing the required nutrients for the growth of microorganisms were used as feed to the SBRs. Experiments included duplicate controls that contained no oxidized test chemicals. The reactor contents were continuously monitored over a 2-week period for MLSS, COD, TOC and the parent compound, when it existed.

2.2.2 Chemical Oxidation Tests

2.2.2.1 Ozonation

The ozonation set-up consisted of an ozone generator (Osmonic OREC V-10), an oxygen generator or a pressurized tank, monitoring devices, and a glass reaction vessel of four liters in volume (Figure 2-1). The reaction vessel, which has openings for ozone inlet, outlet, and for liquid sampling, was operated in semi-batch mode. A glass diffuser was used to sparge ozone gas into the solution. Teflon tubing was used for all the connections from the ozone generator to the reaction vessel. A piston pump with ceramic liner and piston (FMI Model QD) was used to take samples from the reactor at various time intervals. The amount of ozone produced was controlled by changing the power input into the generator. Temperature was maintained at 20 °C ±0.2 in the reactor, which was mixed continuously by a Teflon-coated stirrer at the highest speed using a Fisher magnetic stirrer.

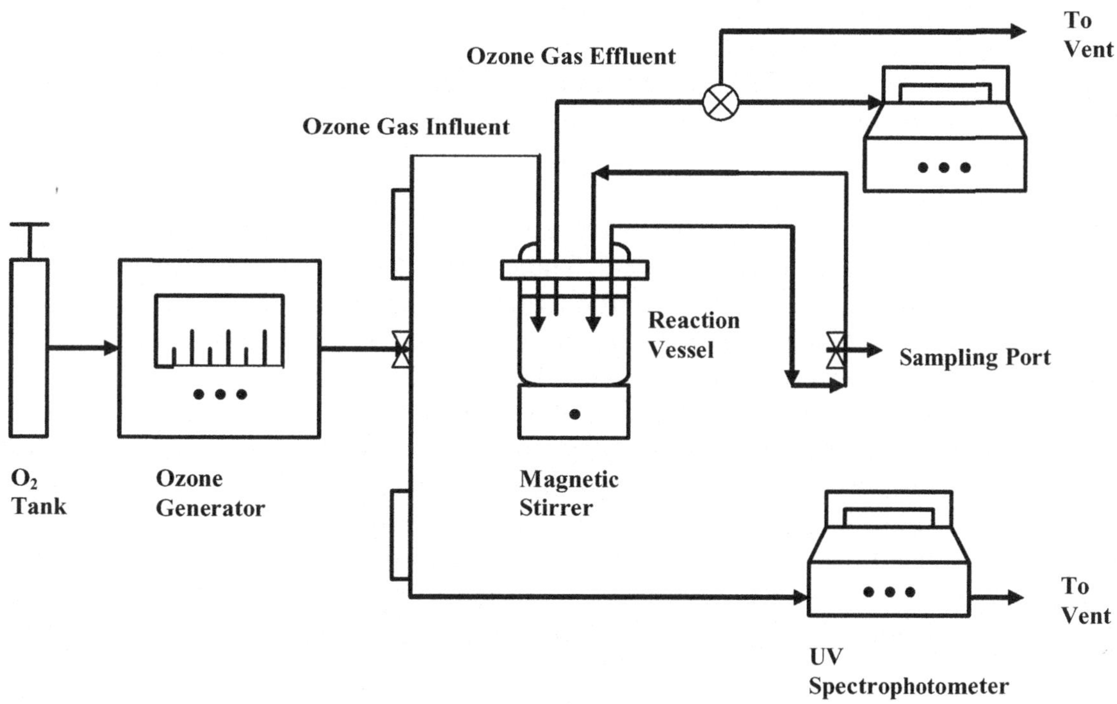

Figure 2.1. Experimental Setup for Ozonation.

The ozone gas concentrations in the influent and effluent were measured by a Shimadzu UV-1601 UV-Visible Spectrophotometer that was calibrated by the potassium iodide method (Standard Methods, 1995). Ozone gas flow was controlled by a gas flow meter that was calibrated by the soap film technique. The gas flow from the ozone generator was diverted to the soap-film column, and the rubber bulb was squeezed to produce a soap-film in the column. Gas flowing through the column moved the film upwards, and the time required for the gas to replace the known volume of air in the column was used to determine the flow rate. A glass diffuser was used to sparge ozone gas into the solution. The solution in the reactor was mixed by a Teflon-coated stirrer at the highest speed using a Fisher magnetic stirrer. Teflon tubing was used for all the connections from the ozone generator to the reaction vessel. A piston pump with ceramic liner and piston (FMI Model QD) was used to take samples from the reactor at various time intervals.

2.2.2.2 Fenton Reaction

A glass reaction vessel of four liters which can accommodate three liters of solution was used for the Fenton experiments. H_2O_2 and $FeCl_3$ or $Fe_2(SO_4)_3$ were either added to the reactor at the beginning of the reaction, or added in stepwise fashion with 10-minute intervals. The pH was adjusted to 3.0 using 1 N or 0.1 N H_2SO_4. During the reaction period samples were taken at 10 minute-intervals. After the reaction was over, NaOH and sodium bicarbonate solutions were added to the reaction mixture to increase the pH to 7, and then the solution was filtered using 0.45 μm membrane filters for separation of the iron precipitate before any measurements. At the end, the oxidized solutions contained 5 mM sodium bicarbonate. A catalase solution of 2,390 units/mg was used to quench any remaining H_2O_2 prior to the inhibition studies. Per vendor instructions, 1 unit of catalase was used to quench 1 μmole of H_2O_2 per minute. The catalase was given up to 10 minutes to react at 25°C. Temperature was maintained at 20±0.2°C in the reactor, which was mixed continuously by a Teflon-coated stirrer at the highest speed using a Fisher magnetic stirrer.

2.2.2.3 UV/H_2O.

The H_2O_2/UV oxidation studies were carried out in a 4-L quartz reactor vessel using the experimetal set-up shown in Figure 2-2. Three liters of test solution was used. The solution was mixed thoroughly before introduction into the reactor. The reactor was operated as a completely mixed batch reactor. Mixing was achieved by circulating the vessel contents at a rate of 200 mL/min by a peristaltic pump. The cylindrical quartz reaction vessel (diameter of 140 mm and a height of 260 mm) was placed inside a Rayonet Photochemical Reactor, equipped with eight low-pressure mercury vapor UV lamps surrounding the reaction vessel. For each test, the reaction vessel was filled with three liters of solution, connected to the circulating pump and exposed to UV light. The total output of the lamps at 254 nm was 5.37×10^{-6} einsteins/L-sec, as determined by actinometric measurement. Samples were taken for analysis utilizing a three-way control valve placed on the circulation. Catalase was used to quench any remaining H_2O_2 prior to the biodegradation and inhibition studies.

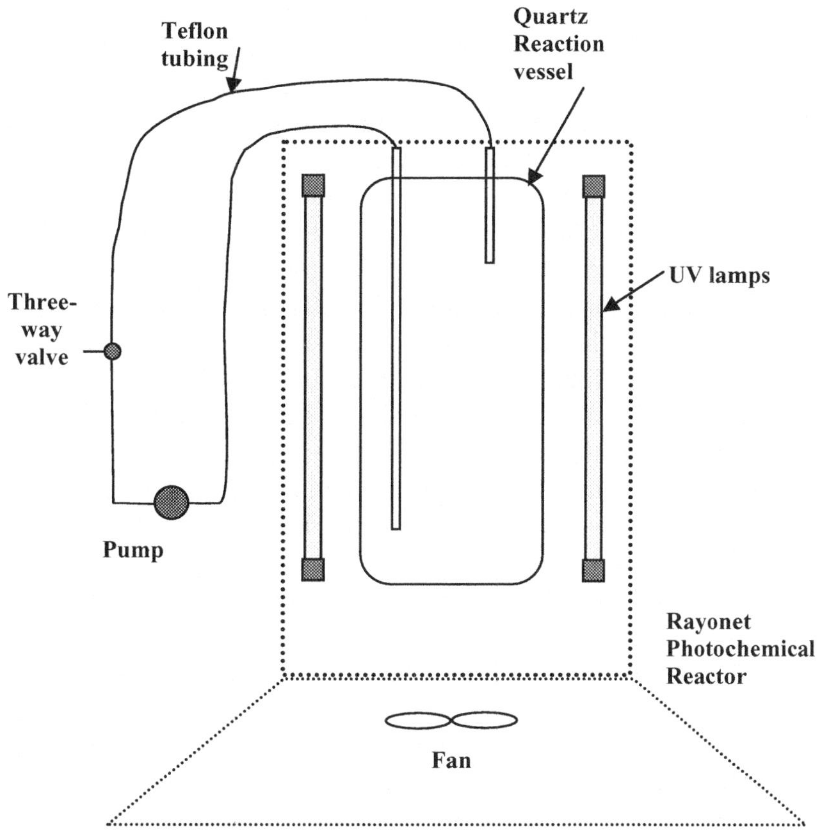

Figure 2-2. UV/H_2O_2 Oxidation Experimental Setup.

2.2.3 Activated Sludge Culture

Heterogeneous microbial populations used in this study were grown in our laboratory in two 5L reactors, both operated in semi-batch mode. The original biomass was obtained from the return line of the return activated sludge (RAS) reactor in the the North City Reclamation Plant, San Diego, CA. The mixed liquor suspended solids (MLSS) concentration in the reactor was adjusted to and maintained at about 3,200 mg/L. The biomass was fed with a mixture of filtered primary effluent and COD supplement solution to adjust the strength of the feed to about 1,500 mg/L as COD. The COD supplement solution was prepared using the modified recipe of Henze (1992). The recipe contained the following chemicals per liter of distilled water: acetic acid, 2.5 mL; butyric acid, 2.5 mL; propionic acid, 2.0 mL; ethanol, 1.0 mL; glutamic acid monosodium salt, 5 g; glucose, 3.75 g; dibasic potassium phosphate, 80 g; monobasic potassium phosphate, 40 g. The primary effluent was collected once a week from the same plant, and the COD supplement solution was prepared in our laboratory. The mixture of filtered primary effluent and COD supplement solution was added to the reactors daily at a volume of 4 liters.

Upon arrival at the San Diego State University (SDSU) laboratories, the primary effluent was filtered through a high-porosity filter paper and then through a glass-fiber filter paper with a pore size of 0.45 μm. The filtrate was refrigerated to slow down the biodegradation of the organic compounds. Prior to its use, the COD concentration of the primary effluent was measured by the closed reflux, colorimetric method (*Standard Methods,* 1995). The strength of the feed was adjusted to 1,500 mg/L as COD by mixing a predetermined amount of primary effluent with the predetermined amount of the COD supplement solution. Twice daily, 2L of the supernatant were withdrawn from the reactors to maintain the desired liquid and solids residence times. The operating parameters of the reactors consisted of sludge residence time of 5 days, food to microorganisms ratio (F/M ratio) of 0.375 day^{-1} and a hydraulic residence time (HRT) of 1.25 days, which is equivalent to an extended aeration mode for an activated sludge system. The reactors were aerated with compressed air (1 L/min), which maintained about 4 mg/L of dissolved oxygen at ambient temperature (22°C±2). Once a week, one liter from each reactor was replaced by fresh RAS from the same plant in order to maintain a diverse microbial community. Prior to experimentation, non-acclimated cultures were grown in our laboratories for at least two weeks. Also, in order to acclimate the microorganisms, one of the two reactors was treated with 0.4 mg/L of DCDE in addition to the mixture of the primary effluent and the COD supplement solution for at least 4 weeks before the experiments were conducted.

A standardized biomass harvesting protocol was used for each experiment. One liter of the biomass was withdrawn from the reactors and the biomass was harvested by centrifugation at 3,500xg for 15 minutes. The pellet was recovered and resuspended in one liter of mineral salt solution (MSS). The MSS was prepared by adding: dibasic potassium phosphate, 86 mg; monobasic potassium phosphate, 217.5 mg; monobasic sodium phosphate, 26.24 mg; ammonium chloride, 20 mg; magnesium sulfate, 22.5 mg; calcium chloride, 27.5 mg; ferric chloride, 0.25 mg; manganese sulfate, 0.04 mg; boric acid, 0.06 mg; zinc sulfate, 0.04 mg; and ammonium molybdate, 0.03 mg to 1 L of distilled water.

After harvesting, the biomass was aerated at an air flow of 1.65 L/min for 2 hours to achieve a stable rate of endogenous respiration during which suspended solids analysis was performed. The biomass was adjusted with MSS to yield 2000 mg/L as MLSS.

2.2.4 Biodegradability Tests

This section describes the specific biodegradation test methodologies used in this research to assess the effects of advanced oxidation on biodegradability of the compounds tested.

2.2.4.1 Short-Term Biodegradation: Test Methodology

The short-term biodegradation test involved the measurement of biomass respiration rate by oxygen probes for a period of 30 minutes. This test was conducted using DCDE over a wide concentration range while keeping the biomass concentration constant at 400 mg/L as suspended solids. This concentration of microorganisms was chosen to yield a significant oxygen decrease within 30 minutes of experimentation, thereby allowing accurate measurement of respiration rates. A modification of the Organization of Economic Cooperation (OECD) Method 209 (OECD, 1987) was followed as the experimental protocol for the short-term tests. The method was based on the measurement of endogenous and exogenous respiration rates with dissolved oxygen (DO) probes. The experiments were conducted in BOD bottles which were able to accept the DO probes (Orion, Model #97-08).

A personal computer, with a Pentium II processor, equipped with a data acquisition board (IO Tech, Model DynaRes-8U) was set up to acquire the DO data during the biodegradation experiments and then to calculate the kinetic constants of the biodegradation rate. The short-term biodegradation tests were performed with both non-acclimated and acclimated microorganisms. The biomass concentration was adjusted to 2000 mg/L as suspended solids with MSS after harvesting and placed in 60 mL aliquots to each BOD bottle so that in the final volume of 300 mL, the concentration was consistent at 400 mg/L as suspended solids. Samples of oxidized DCDE or MTBE solutions were then added to the BOD bottles in different dilutions (calculations are presented in the tables together with the results), and the volumes were increased to 300 mL with distilled water. Yeast extract (0.15 mg/L) was added to each BOD bottle before starting the experiments to provide the necessary macro nutrient for the microorganisms. The oxidized and non-oxidized solutions of DCDE or MTBE were tested in triplicate. The duration of the experiments were kept at 30 minutes in order to prevent further acclimation of microorganisms, as well as to prevent complete depletion of oxygen. Using DO probes attached to a highly sensitive data acquisition system allowed the collection of data at a very high speed: one data point per second per bottle. The respiration rate of the microorganisms in each bottle was determined by measuring the oxygen concentration as a function of time and by calculating the corresponding slopes using Microsoft Excel program. Figure 2-3 shows typical raw data obtained for a DCDE experiment at different dilutions, where time is in real minutes.

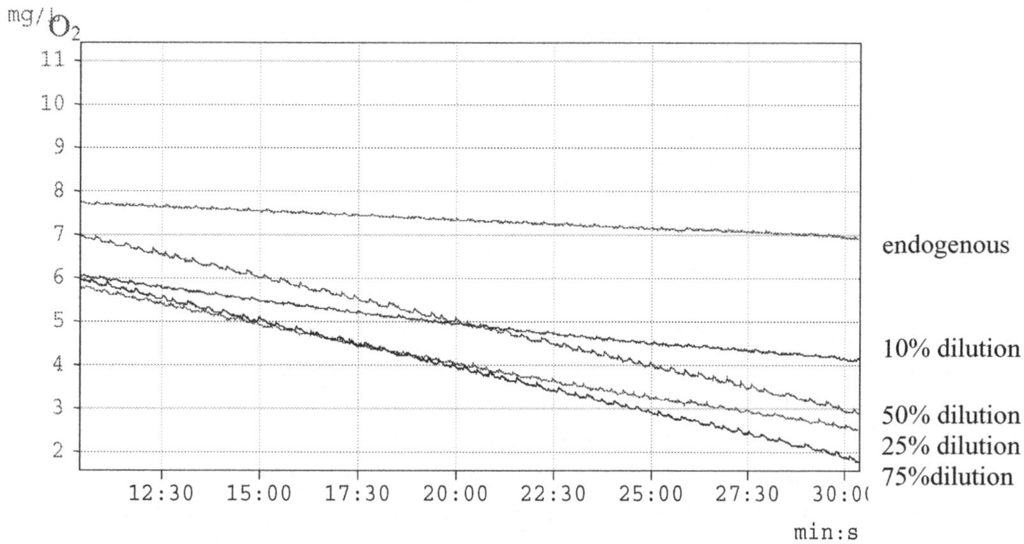

Figure 2-3. Respiration Rates of Microorganisms with Different Dilutions of DCDE After 75% Oxidation by Ozone.

The endogenous rate of respiration of microorganisms was obtained by running a control experiment with microorganisms alone. The net exogenous respiration rates for DCDE or MTBE solutions were obtained by subtracting the endogenous rate from the observed exogenous respiration rates. The net exogenous respiration rates that are directly proportional to the substrate removal rate were substituted in the Monod equation to determine the specific substrate removal rate constants:

$$-\frac{dS}{dt} = q = \frac{q_{max} \cdot S}{S + K_S}$$

Non-parent TOC resulting from oxidation was used as the biodegradable substrate concentration (S) for parameter estimation. Non-parent TOC for this purpose was defined as the TOC contributed by the chemical oxidation products, and it was estimated by subtracting the TOC attributable to the parent compound from the total TOC. For example, for DCDE, non-parent TOC=npTOC = total TOC – $0.34C_{DCDE}$, where total TOC is the measured TOC value, C_{DCDE} is the measured concentration of DCDE, and 0.34 is the theoretical amount of TOC per mg of DCDE. The constants, q_{max} and K_S were determined through a curve fitting method used by Sigma Plot software. The method involves iterative non-linear curve fitting by minimizing least squares errors. The curve fitting and statistical significance tests were performed by using the software for the 95% confidence level. Comparison of the kinetic constants obtained for the oxidized samples and the non-oxidized samples allowed the assessment of the extent of improvement of biodegradation.

2.2.4.2 Long-Term Biodegradation: Test Methodology

The long-term respirometry experiments were conducted over a period up to 14 days to obtain information on biodegradability and toxicity of the by-products of oxidation and of the parent compound. These data demonstrated whether prolonged exposure would prompt microorganisms to develop the necessary enzymes for degradation of the parent compounds and/or the oxidation by-products, or would result in toxicity and killing of the microorganisms.

A 10-channel Micro Oxymax Respirometer (Columbus Instruments International Corporation, Columbus, OH), equipped with oxygen and carbon dioxide sensors and a computer interface, was used to assess the improvement in the biodegradability of the parent compounds after 0%, 25%, 50%, 75%, and 100% oxidation of the parent compounds by ozone or Fenton reagent. The respirometric technique involved placing the oxidized or non-oxidized solutions with microbial inocula into sealed 250 mL bottles and continuously measuring oxygen consumption and the production of carbon dioxide by the bacteria as they grow under batch conditions while degrading the substrate. The respirometer was used for continuous measurement of oxygen and carbon dioxide in the gas phase up to 14 days. The process involved automatic gas withdrawal from the test chambers, which was pumped through the gas sensors and then returned to the test chambers. The percent oxygen and carbon dioxide gas levels of the test chamber environment were measured every three hours and the changes in the levels were used to compute oxygen consumption and carbon dioxide production. Micro Oxymax software stored the data obtained during the experimental period directly to a hard disc.

Respirometric studies were conducted with acclimated and non-acclimated cultures of microorganisms harvested from the activated sludge reactors as before. The appropriate amount of biomass was placed into the respirometer bottles to result in 400 mg/L of biomass as suspended solids along with the oxidized solutions diluted by 50%. The bottles were capped and the experiments started. All the tests were run in duplicates.

The experiments included three levels of controls, each with duplicates: 1) test with no microorganisms, 2) test with microorganisms and a known amount of non-oxidized parent compound (no oxidation), and 3) test with microorganisms alone (endogenous).

2.2.4.3 Mid-Term Biodegradation: Test Methodology

These experiments were designed to present data in terms of degradation of the biodegradable organic matter, as measured by the changes in TOC and COD content of the solutions as a function of time. The total amount of oxidation by-products that were biodegradable within six hours was then determined and compared among different treatments. Six hours was used as the time span because 1) the rate of biodegradation as measured by COD or TOC change leveled off beyond six hours, and 2) six hours is a typical hydraulic residence time in wastewater treatment plants.

The amount of TOC and COD due to the parent compound was subtracted from the total measured concentrations of TOC and COD, and the remaining amounts of TOC and COD were attributed to the oxidation products. For example, the theoretical COD of 1 mg of DCDE is 160/143 = 1.12 mg. Hence, the COD contributed by the oxidation products of DCDE, i.e., "the non-parent COD", was estimated as follows: npCOD = (Total COD – $1.123C_{DCDE}$). The npTOC was defined earlier.

A different experimental system was set up to conduct the mid-term biodegradation tests (Figure 2-4). These experiments were conducted in duplicates in 1 L glass bottles open to the atmosphere. Air was supplied with glass diffusers with a flow rate of 1.65 L/min. The concentration of microorganisms, measured as MLSS, was adjusted to 400 mg/L, with dilution of all samples by 50% once the solution containing the parent compound was added.

The mid-term experiments were performed using oxidized DCDE and MTBE and acclimated or non-acclimated cultures. Over 24 hours, samples (20 mL) were withdrawn every 1.5 hours, filtered directly into 20 mL vials through 0.2 μm pore filter papers, and the filtrate immediately analyzed for the concentrations of the parent compounds. The remaining filtrate was acidified with H_2SO_4 and stored at 4°C. The measurements of suspended solids, COD and TOC were performed within 24 hours.

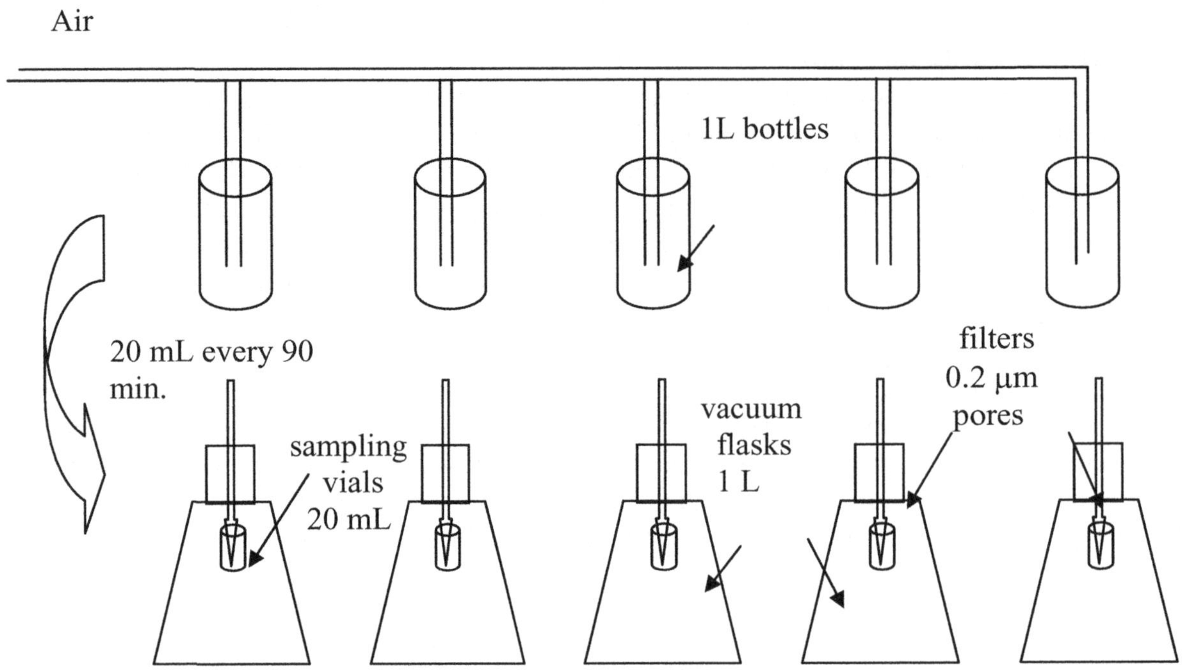

Figure 2-4. Schematic of Experimental Setup for Mid-Term Biodegradation Tests.

2.2.5 Inhibition Tests: Test Methodology

A modified version of the OECD Method 209 was used to monitor respiration rates and quantify inhibition caused by the test chemicals and their oxidation byproducts. For each test, six samples were poured into BOD bottles and continuously stirred with Teflon-coated magnetic stir bars. DO probes (Orion, Model #97-08) were fitted securely with plastic inserts through the openings of the BOD bottles to prevent air entering or escaping. The plastic inserts came with the DO probe equipment and are designed specifically to fit the DO probe and a standard BOD bottle. Cables from the DO probes led to a data acquisition board (IO Tech, Model DynaRes-8U) that transferred data for dissolved oxygen concentrations to a personal computer with a Pentium II processor. Prior to each experiment, DO probes were calibrated at their zero point and for air concentration (approximately 6.5 mg/L) using designated dials on the probes.

In each experiment, one BOD bottle was used to measure the endogenous respiration (no acetate). Two additional bottles were used to measure the exogenous respiration (acetate) and three more to measure inhibited exogenous respiration (acetate plus inhibitor). The endogenous respiration rate was calculated from the measurement of oxygen depletion over time by activated sludge in its "resting state". Since the microorganisms at this stage are without food, they are not very active. Hence, their oxygen demand is minimal. The exogenous respiration rate was calculated from the rate of oxygen depletion in the presence of a readily biodegradable substance. Since the microorganisms are actively biodegrading the substrate, their oxygen demand is high and the concentration of dissolved oxygen in a closed system drops dramatically. In these studies, acetate was used as a readily biodegradable substance. Lastly, inhibited exogenous respiration rate represents the decrease in the rate of oxygen consumption over time in a system with microorganisms and acetate in the presence of a potentially inhibitory compound or combination of compounds.

To check the inhibition of the test chemicals, four experiments were performed with concentrations of the test chemicals at 25, 50, 75 and 100 mg/L, an activated sludge concentration of 400 mg/L, and an acetate concentration of 20 mg/L. The respiration rates were monitored for 15–20 minutes.

To check the inhibition of the oxidized solutions, six experiments were performed after 25%, 50%, 75%, 100%, 150%, and 200% oxidation of the test chemical, using 400 mg/L activated sludge and 20 mg/L acetate. For ozonation and UV/peroxide treatment, the oxidation levels of 150 and 200% were obtained by allowing the reaction to continue for periods of 1.5 and 2.0 times the time needed for 100% oxidation of the test chemical. For Fenton treatment, the oxidation levels of 150% and 200% were obtained by adding chemicals that are 1.5 and 2.0 times the amount needed for 100% oxidation of the test compounds. The initial concentration of the test chemicals in solutions prior to oxidation was approximately 200 mg/L for all experiments. The level of oxidation achieved was within 5% of the desired level. Test samples were capped prior to the inhibition studies to avoid volatilization of chemicals. All experiments were repeated using acclimated and non-acclimated cultures.

2.2.6 SBR Tests

A schematic of the experimental set-up to simulate SBRs in this study is shown in Figure 2-5. A set-up of only one reactor is presented in the figure for clarity; however, six reactors were used simultaneously in these studies.

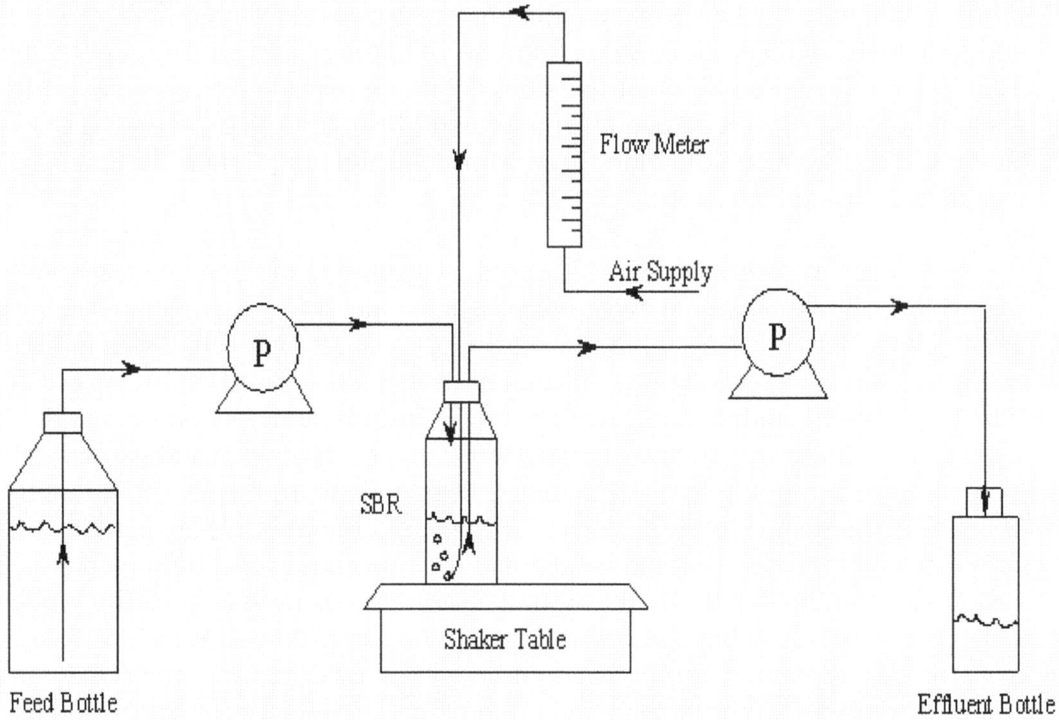

Figure 2-5. Experimental Setup for SBR Studies.

The reactors used in this study consisted of 500 mL Pyrex bottles utilizing a Kontes delivery cap (#953930) for transfer of air and fluid. The ports in the delivery cap were used for air introduction, air effluent, feed introduction, and supernatant effluent. A manifold containing six flow meters (Cole-Parmer, NO12-10ST) was used to distribute 400 mL/min of compressed air to each reactor. A glass tube was inserted through the delivery cap to the desired level of the reactor (250 mL), so that the reactor volume would be withdrawn to 250 mL at the end of each cycle. Peristaltic pumps were used to introduce feed into the reactors and to withdraw supernatant. The effluent lines from each reactor were distributed to six plastic bottles for effluent collection. Each reactor was placed on a shaker table (Labline Instruments #3518) and kept at a speed of 125-135 rpm. The feed bottles were 1 L Pyrex bottles and were connected to the feed tubing by a brass barb (1/8"). Polytetrafluoroethylene (PTFE) tubing was used for all connections from the feed bottles to the reactors to the effluent bottles.

The solutions prepared in this study included Syntho solution, the supplement solution, and oxidized solutions (Table 2-1).

Table 2-1. Syntho Solution Constituents[1].

Carbon Source (mg/L)	Nitrogen Source (mg/L)	Phosphorous Source (mg/L)	Sewage Simulation (mg/L)
Peptone 15	Urea 75	$MgHPO_4{\cdot}3H_2O$ 25 **Replaced with** KH_2PO_4	LAS[2] 10
Na-acetate 120	NH_4Cl 11	$K_3PO_4{\cdot}H_2O$ 20 **Replaced with** Na_2HPO_4	AE[3] 10
Dry Meat Extract 15	Uric Acid 9		Lyophilized activated sludge 50 **Not Used**
Glycerol 40			Diatomaceous Earth 10
Potato Starch 50			Diet Fibers 80 **Not Used**
Low Fat Milk Powder 120			
Other Minerals and Trace Metals: $CaCl_2$ 5 mg/L **Replaced with $CaCl_2{\cdot}2H_2O$** $MnSO_4{\cdot}H_2O$ 100 μg/L $CuCl_2{\cdot}2H_2O$ 480 μg/L **Replaced with $CuSO_4$** $CoCl_2{\cdot}6H_2O$ 50 μg/L K_2MoO_4 20 μg/L **Replaced with $(NH_4)_6Mo_7O_{24}{\cdot}4H_2O$** $ZnCl_2$ 180 μg/L $FeSO_4{\cdot}7H_2O$ 10 mg/L $NiSO_4{\cdot}6H_2O$ 300 μg/L $NaHCO_3$ 25 mg/L EDTA 0.22 mg/L $Cr(NO_3)_3{\cdot}9H_2O$ 680 μg/L			

[1] Lyophilized activated sludge (AS) was added to real domestic wastewater to maintain a broad microbial diversity in the source journal. In our experiments, wastewater was added daily to maintain microorganism diversity, so the lyophilized AS was deemed unnecessary. Also diet fibers were said to help the settling characteristics. There was no difficulty in the sludge settling in these experiments so the fiber was omitted so that feed tubing would not be clogged.
[2]Linear Alkylbenzene Sulphonate (LAS)
[3]Alcohol Ethoxylate (AE)

The feed solution was the final desired combination of these solutions. Syntho is a synthetic sewage solution that was developed by Boeije et al. (1999) for routine use in laboratory-scale sewage treatment simulation and biodegradation tests. Results from their test showed that Syntho is a close approximation to real sewage for laboratory purposes. Syntho was used because of the ease of preparation and to avoid microorganism growth in the feed lines that would occur from real sewage. The supplement solution was used to increase the COD of the feed solution.

Each feed bottle with the Syntho and supplement solution (if used) was autoclaved for 30 minutes in order to prevent contamination. Oxidized and spiked solutions were added subsequent to autoclaving. Initially the reactors were prepared by adding 100 mL of mineral buffer solution, 5 mL of supplement solution, 20 mL of filtered sewage, and 125 mL of returned activated sludge from the North City Reclamation Plant in San Diego, CA. A retention time of nine hours required a batch volume of 30 mL of feed solution to the reactors every hour. Each reactor had a volume of 250 mL before each batch addition. The working batch volume was 280 mL. The reactors simulated SBRs with each cycle consisting of 60 minutes of operation as follows: 3-minute fill period, 52-minute react period, 2-minute settle period, and 3-minute decant period.

The experiments were controlled via a personal computer with a Pentium II processor equipped with a data acquisition board (IO Tech, Model DynaRes-8U). The system was operated on a timer via a software program (Dasylab). At the beginning of each cycle, the air and shaker table were initiated and operated until the settle and decant period. Also, the influent pump was initiated to introduce 30 mL of feed solution into each reactor. At the end of the react cycle, the reactors were allowed to settle, and then the effluent pump was started to remove the supernatant and waste some activated sludge. After supernatant was removed, the effluent pump shut off, and the cycle was repeated.

Each reactor was allowed to achieve equilibrium for 1-3 days utilizing only Syntho solution as the feed. Oxidized chemical was added to the respective reactors on the second or third day after the contents of the reactors was mixed and evenly redistributed to each reactor. For the figures presented in the results section, day zero represents the day chemicals were added (unless otherwise stated).

Reactors 1 and 2 were the controls and received no chemical, whereas reactors 3 and 4 received 50%–75% oxidized chemical, and reactors 5 and 6 received 85%–100% oxidized chemical (unless otherwise stated). The COD and TOC of the oxidized chemical and the Syntho solution were measured to determine the contribution of COD and TOC from the by-products and from the Syntho solution for each experiment. Two different types of feed solutions were used: (1) a solution with a COD and TOC of 1,100-1,500 mg/L and 450–550 mg/L, respectively (referred to as "concentrated feed"), and (2) a solution with a COD and TOC of 300–350 mg/L and 100–115 mg/L, respectively (referred to as "dilute feed").

The reactors were allowed to run 10–14 days. Throughout the experiment, 50 mL of filtered sewage was added to each reactor every day to maintain a diverse culture of microorganisms. Influent, effluent, and reactor samples were taken every two days starting after the reactors were mixed. These samples were tested for COD, TOC, chemical concentration, and suspended solids (SS).

2.2.7 Analytical Methods

The concentration of DCDE was measured by a Shimadzu Gas Chromatograph equipped with a flame ionization detector. Aqueous samples were injected directly on a Nukol capillary column (length 15 m, internal diameter 0.53 mm and film thickness 0.5 μm), which was obtained from Supelco (Bellefonte, PA). Helium was used as a carrier gas at 15 mL/min, while hydrogen and air were used to fuel the detector. The injector and detector were operated at the temperatures of 150°C and 250°C, respectively. The initial column temperature was 104°C, and it was raised to 200°C at a rate of 8°C/min.

Aldehydes were derivatized in aqueous-phase by using o- (2,3,4,5,6-pentaflurobenzyl)-hydroxylamine hydrochloride (PFBHA-HCl) as a derivatizing agent, according to the Standard Methods (APHA, 1995). The derivatized aldehydes were then analyzed by a Hewlett Packard Gas Chromatograph equipped with an electron capture detector. The procedure used here was originally described by Yamada and Somiya (1989).

The carboxylic acids were analyzed using a Dionex Ion Chromatography. The ion chromatographic method was developed recently by the Metropolitan Water District of Southern California (Kuo, 1998). The method requires the use of an Ionpack AS11 analytical column, an Ionpack AG11 guard column, and an anion-trap column, which were obtained from Dionex. An isocratic and gradient-stage separation mode was used in the process. The degassed ultra pure water and 0.1 mM and 1 mM NaOH solutions were used as working eluents.

For identification of other intermediate products, a Hewlett Packard 6890 gas chromatograph with Micro-ECD detector, HP-7683 auto injector and Chem workstation. This GC is connected with HP-5793 Mass Selective Spectrometer and equipped with 59864B ionization gauge controller and search capabilities through the NIST search library.

The TOC concentration was obtained by subtracting the measured concentration of inorganic carbon (IC) from the measured concentration of total carbon (TC). Both TC and IC were measured using TOC Analyzer 5000A (Shimadzu Corporation, Kyoto, Japan). TC and IC are measured by the same principle by measuring the amount of CO_2, which is detected by non-dispersive infrared gas analyzer. TC samples were combusted, and IC was acidified to decompose to become CO_2.

The COD of the samples was measured by the closed reflux, colorimetric method (Standard Methods, 5220-D).

Total suspended solids in samples were measured by the gravimetric method (Standard Methods, 2540-D).

The concentration of H_2O_2 in stock solution was measured by the Permanganate Method (Kolthoff et al., 1969). The titanium (lV) chloride method (Boltz and Howell, 1978) was used to determine the concentration of unreacted H_2O_2 in reaction solutions.

Ozone concentration in the influent and effluent gas was measured by a Shimadzu UV-visible spectrophotometer at the wavelength of 258 nm. The gas was passed through a flow-through cell that had 1 cm path-length. The potassium iodide method was used to calibrate the spectrophotometer (Standard Methods, 1995). The liquid phase ozone concentration was determined by the indigo method (Bader and Hoigne, 1981).

2.2.8 Chemicals

The chemicals used in this research are listed in Table 2-2 with the list of vendor, lot number, grade, and purity.

Table 2-2. List of Chemicals Used in this Research.

Chemical	Formula	CAS #	Vendor	Grade	Purity	Lot #
Acetic acid	CH_3COOH	64-19-7	J.T.Baker	Analytical	99.6	M23833
Ammonium Chloride	NH_4Cl	12125-02-9	Mallinckrodt	Analytical	99.99%	RHY
Ammonium Molybdate	$(NH_4)_6Mo_7O_{24}\cdot 4H_2O$	12054-85-2	Fisher	A.C.S. Reagent	99%	761020
Butyric Acid	C_3H_7COOH	107-92-6	Alfa Aesar	Analytical	99%	A07H18
Calcium Chloride	$CaCl_2\cdot 2H_2O$	10035-04-8	Aldrich	Analytical	>98%	12210JR
Calsoft	LAE	F-90	Pilot Chemical Co.		90%	70104C00112 2
Catalase		9001-05-2	Sigma		2390 units/µg	128H7250
Chromium Nitrate	$Cr(NO_3)_3\cdot 9H_2O$	7789-02-8	Arcos	Analytical	99%	A011737501
Cobalt Chloride	$CoCl\cdot 6H_2O$	7791-13-1	Arcos	Analytical	99.9%	A0140019
Copper Sulfate	$CuSO_4$	7758-99-8	Aldrich	Analytical	>98%	10931CR
Dibasic Potassium Phosphate	K_2HPO_4	7768-11-4	J.T.Baker	Analytical	100.4%	31048
Dichloro diethyl ether	$C_4H_8OCl_2$	111-44-4	Sigma-Aldrich	Analytical	99%	04219HS
Dodecyl-benzene sulfonic acid	$C_{18}H_{29}NaO_3S$	25155-30-0	Fluka		80%	44200
EDTA		EX-539	Clarkson	Analytical		
Ethanol	C_2H_6OH	64-17-5	Aldrich	Analytical	99.5%	KU07321EU
Ferric Chloride	$FeCl_3$	7705-08-0	Aldrich	Analytical	97%	13812ER
Ferric Sulfate	$FeSO_4\cdot 7H_2O$	10028-22-5	Sigma	Analytical	23% Fe	80K2004
Glucose	$C_6H_{12}O_6$	492-62-6	Aldrich	Analytical	100%	06129MR
Glutamic Acid, Mono-sodium Salt	$C_5H_8NO_4Na$ x H_20	142-47-2	Acros	Analytical	99%	A010720701
Glycerol	$C_3H_8O_3$	56-81-5	Aldrich	A.C.S. Reagent	>99.5%	DS07912TR
Hydrogen Peroxide	H_2O_2	7722-84-1	Aldrich	Sol'n in Water	30 wt.%	10413AS
Mercuric Sulfate	$HgSO_4$	7783-35-9	Mallinckrodt	Analytical	99.9%	WRML
Methyl *tert*-butyl ether	$(CH_3)_3COCH_3$	1634-04-4	Aldrich	HPLC	99.8%	KU03335KU
Nickel Sulfate	$NiSO_4\cdot 6H_2O$	10101-97-0	Fisher	A.C.S. Reagent	99.99%	004983
Peptone		11843	Becton	Analytical		1000K9 DJDO
Potassium Chloride	KCl	7447-40-7	Aldrich	Analytical	99%	03718PQ

Potassium Dichromate	$K_2Cr_2O_7$	7778-50-9	Mallinckrodt	Analytical	99.9%	WRMR
Potassium Hydrogen Phthalate	2-(HO_2C) $C_6H_4CO_2K$	877-24-7	Mallinckrodt	Analytical	99.95%	04629KN
Propionic Acid	$CH_3CH_2CO_2H$	79-09-4	Alfa Aesar	Analytical	99%	I03103
Silver Sulfate	Ag_2SO_4	10294-26-5	Mallinckrodt	Analytical	99.95%	PTY
Sodium Acetate	CH_3COONa	127-09-3	J.T. Baker	Analytical	99%	32368
Sodium Hydroxide	NaOH	1310-73-2	J.T. Baker	A.C.S. Reagent	98.7%	L49935
Sodium bicarbonate	$NaHCO_3$	144-55-8	Aldrich	Analytical	99.7%	08818JW
Sodium Chloride	NaCl	7647-14-5	Fisher	Analytical	99.9%	771808
Sodium Phosphate	Na_2HPO_4	7558-79-4	Fisher	A.C.S. Reagent	>99%	772503
Sulfuric Acid	H_2SO_4	7664-93-9	J.T. Baker	A.C.S. Reagent	96.4%	G32046
Urea	$(NH_2)_2CO$	57-13-6	Mallinckrodt	Analytical	99.9%	8648
Uric acid	$C_5H_4N_4O_3$	69-93-2	Arcos	Analytical	>99%	79333/1
Zinc Chloride	$ZnCl_2$	7646-85-7	J.T. Baker	Analytical	99.9%	L47338

CHAPTER 3.0

CHEMICAL OXIDATION TESTS

3.1 Ozonation

3.1.1 Characterization of Ozonation System

Experiments were conducted to evaluate the mass transfer rate in the ozone reactor. The ozone concentration in the influent gas was adjusted to 29.5 mg/L (1.5% by volume), and the gas flow rate was set at 1.5 L/min at room temperature of about 20°C. Three liters of pure water at pH 3, where ozone decomposition is negligible, was ozonated over a time period. The results are presented in Figures 3-1 and 3-2. Within about 10 minutes, the effluent ozone concentration in gas reached the same value as the influent concentration. The steady-state concentration achieved in pure water under these conditions was measured as 8.7 mg/L. This value was taken as the $[O_3]^*$, i.e., the equilibrium concentration under the experimental conditions. The Henry's Law constant for ozone as the ratio of ozone concentration in gas divided by ozone concentration in liquid in this case came out to 3.4 (dimensionless), and it agrees with the literature values. The slope of the straight line in Figure 3-3 provides the mass transfer coefficient for ozone (k_La) as 19 hr^{-1}. This information was used for calculation of ozone requirements.

3.1.2 Ozonation of DCDE

Ozonation of DCDE was performed under the experimental conditions of 1.5 L/min ozone gas flow rate, 29.5 mg/L of influent ozone gas concentration, and a liquid volume of 3 L. The solutions had a pH of 7 and 5 mM bicarbonate. Solutions containing two different initial concentrations of DCDE were ozonated for 90 minutes: high-level DCDE — 950 mg/L (6.6 mM) and low-level DCDE — 153 mg/L (1.07 mM).

During the 90-minute period, 37% of 6.6 mM of DCDE and 85% of 1.07 mM DCDE were removed, as shown in Figure 3-4. The data points falling on a straight line on a log-scale show that the reaction followed first order kinetics for DCDE removal.

To quantify the removal of DCDE by stripping due to gas sparging, another set of experiments was conducted under the same influent gas flow rate and a 1.- mM initial DCDE concentration. Oxygen was sparged in place of ozone into the reaction vessel. During 90 minutes of sparging, no removal of DCDE was observed as expected because DCDE has a very low volatility (Henry's Law constant = 1.3×10^{-5} atm-m^3/mol).

Ozone in the influent and effluent gas observed as a function of time during the reaction is presented in Figure 3-5 for both high and low levels of DCDE. The mass-transfer coefficient measured in the absence of DCDE (19 hr^{-1}) together with the data on influent and effluent ozone gas concentration measured in the presence of DCDE allowed making a mass balance for ozone. This calculation indicated that 2.6 moles of ozone were consumed per mole of DCDE destroyed for the 6.6 mM DCDE solution. This stoichiometric number was 3.6 for 1.07 mM DCDE solution.

During ozonation of the DCDE solutions, dissolved ozone concentration in water peaked a little below 6.0 mg/L, and after an initial fluctuation, it continued to decline during the reaction, as shown in Figure 3-6. This observation indicates that rapid consumption of ozone occurred in the presence of DCDE compared to the saturation ozone concentration of 8 mg/L that is expected in water. Because the reaction between DCDE and ozone is apparently very slow, observation of lower dissolved ozone concentration in the solution supports the case for the catalyzed decomposition of ozone by accumulating reaction products, such as formic acid (Can and Gurol, 2003). According to the authors, the reaction of formic acid with ozone produces hydroxyl radical, •OH, which reacts with organic compounds at diffusion-limited rates.

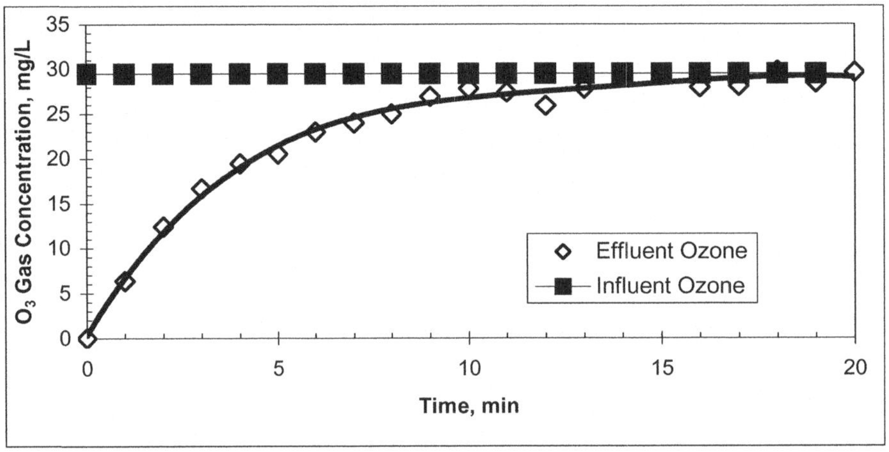

Figure 3-1. Influent and Effluent Ozone Gas Concentration at pH 3.

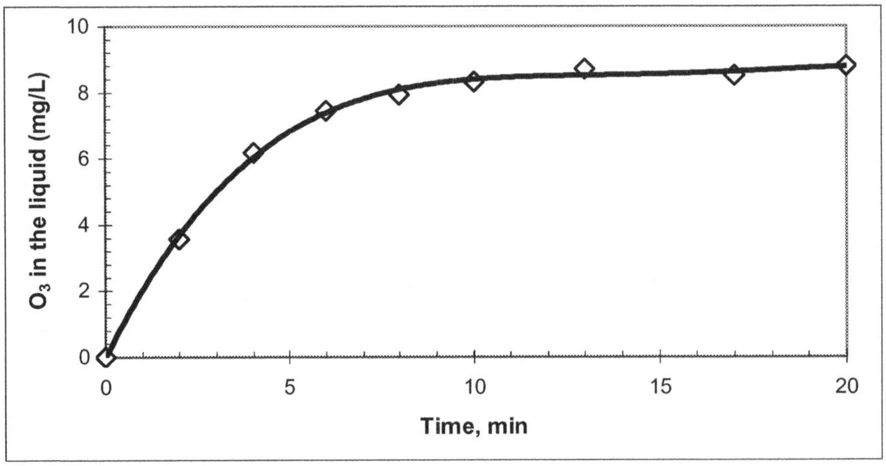

Figure 3-2. Accumulation of Ozone in Solution at pH 3.

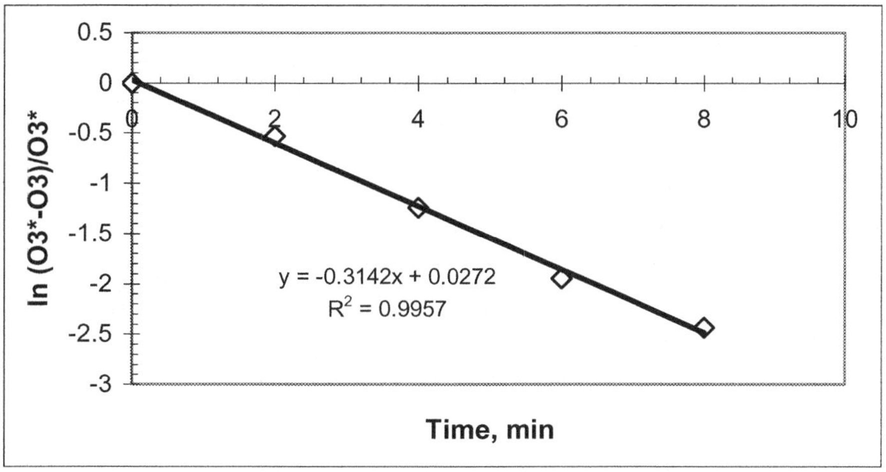

Figure 3-3. Determination of the Mass Transfer Coefficient, k_La.

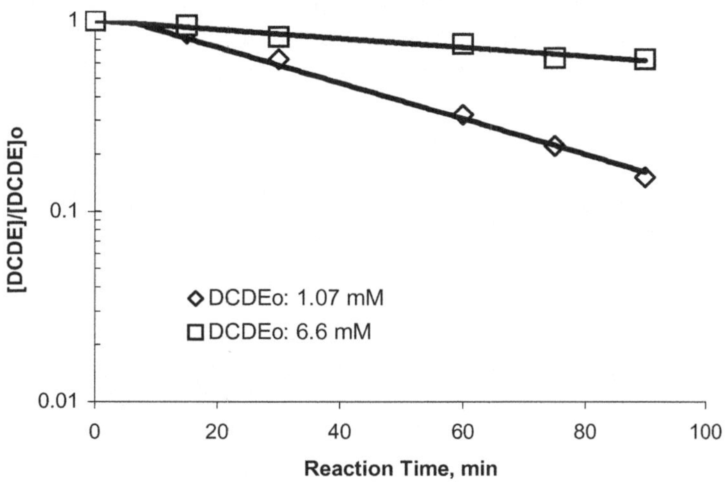

Figure 3-4. Oxidation of DCDE by Ozone.

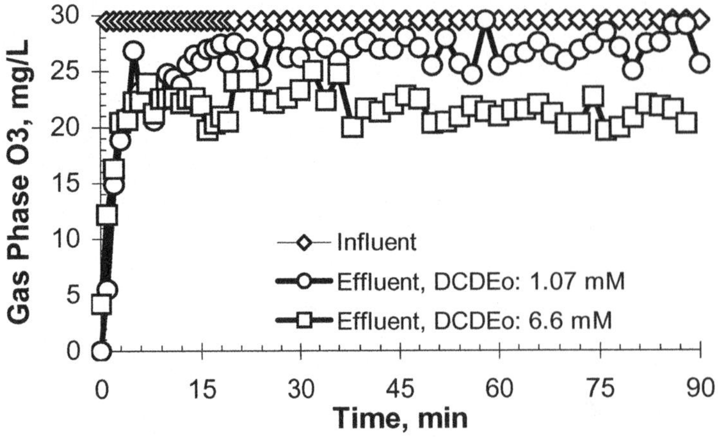

Figure 3-5. Influent and Effluent Ozone Gas Concentrations During DCDE Oxidation.

The role of •OH in ozonation of DCDE was further evaluated by conducting parallel reactions by addition of tertiary butyl alcohol (TBA) as an •OH scavenger. TBA is known to react with •OH by a rate constant of 4.2×10^8-7.6×10^8 M^{-1} s^{-1} (Buxton et. al., 1988). No information is available in the literature on the rate constant of the reaction between DCDE and •OH. Therefore, different amounts of TBA concentrations, i.e., 0.2 mM, 4 mM and 16 mM, were added to quench the •OH and reduce its reaction with DCDE. The removal rates of DCDE in the presence and the absence of TBA are presented in Figure 3-7 for an initial DCDE concentration of 1.07 mM**.** The reaction rate of DCDE was suppressed significantly when the •OH concentration was reduced by TBA, and there was no DCDE removal at all when all •OH was quenched by excessive amounts of TBA. Since ozone does not react with TBA, this observation confirms that DCDE reacted primarily by •OH, and the direct reaction of ozone with DCDE was negligibly slow. Li et al. (1995) also reported that •OH was predominant during oxidation of DCDE by UV/H_2O_2.

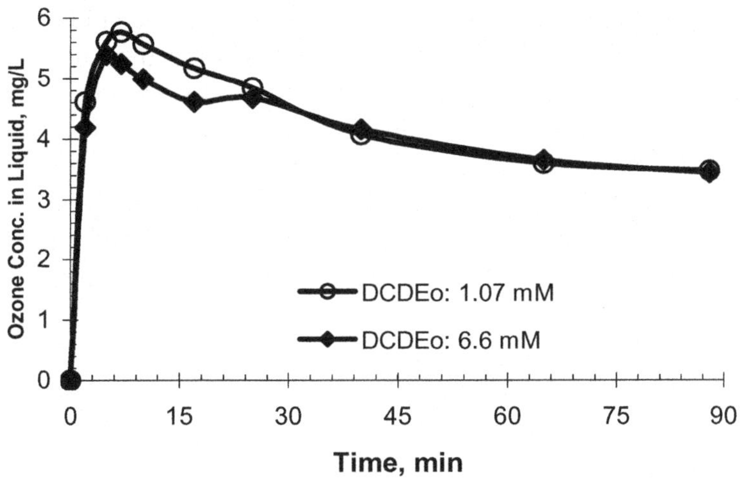

Figure 3-6. Ozone Concentration in Water During DCDE Oxidation.

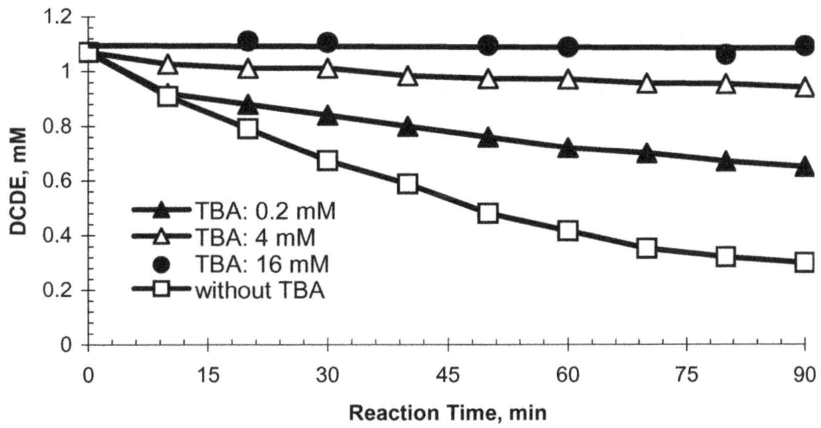

Figure 3-7. Ozonation of DCDE with and without TBA.

3.1.3 Ozonation of MTBE

Since MTBE is a relatively volatile compound, its removal due to stripping alone was first checked in the absence of ozone. In these experiments oxygen rather than ozone was bubbled into the reactor. Within 90 minutes of oxygen bubbling, MTBE was removed by 70% and 46% at flow rates of 1.5 L/min and 0.5 L/min, respectively (Figure 3-8). The mass-transfer coefficient for MTBE at the lower flow rate was half (0.0068 min^{-1}) of the value at the higher flow rate (0.0141 min^{-1}).

Hence, the ozonation experiments for MTBE were conducted at the lower flow rate, 0.5 L/min, to keep stripping of MTBE down. Removal of MTBE during ozonation is presented in Figure 3-9 as a function of time at the 0.5 L/min flow rate. The amount of MTBE removed by ozone oxidation was estimated by subtracting the amount stripped from the total removed. It is clear from these results that MTBE was largely removed by physical stripping rather than chemical oxidation, and therefore ozonation was decided not to be an effective method for oxidation of MTBE. A paper was presented at the AWWA Annual Conference in Washington, D.C., in June 2001 on MTBE removal by chemical oxidation methods, which provided a mechanistic evaluation of MTBE removal by hydroxyl radicals produced during ozonation (Gurol et al., 2001). The contribution of WERF to this study was acknowledged.

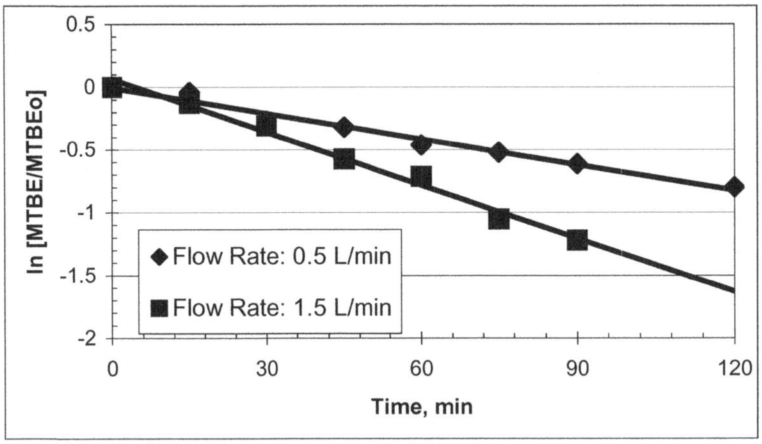

Figure 3-8. Removal of MTBE by Stripping During Oxygenation.

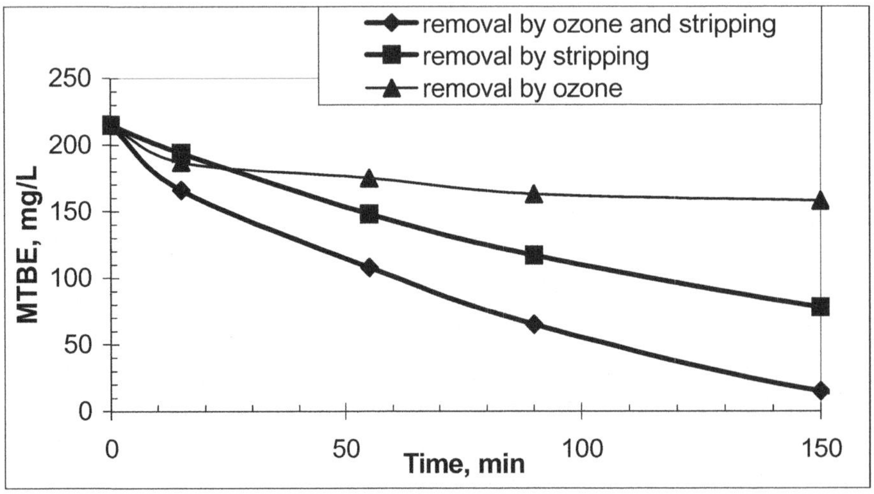

Figure 3-9. Removal of MTBE by Ozonation and Stripping.

Nevertheless, the ozonation system during MTBE oxidation was fully characterized in terms of influent and effluent ozone concentrations in the gas phase and the ozone concentration in liquid, as shown in Figures 3-10 and 3-11.

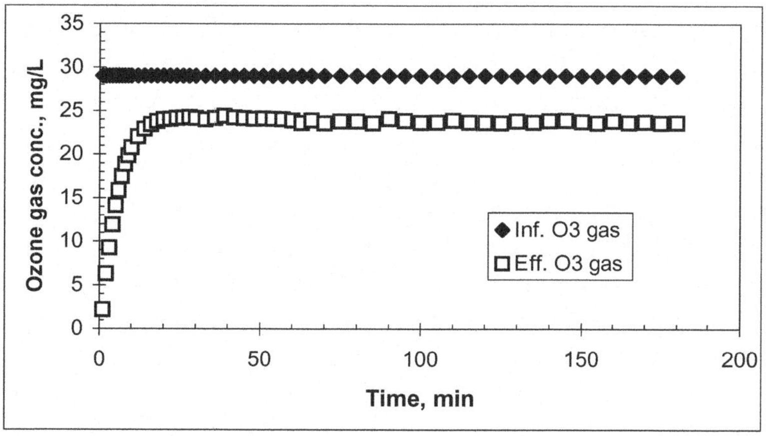

Figure 3-10. Ozone Influent and Effluent Gas Concentration During MTBE Ozonation.

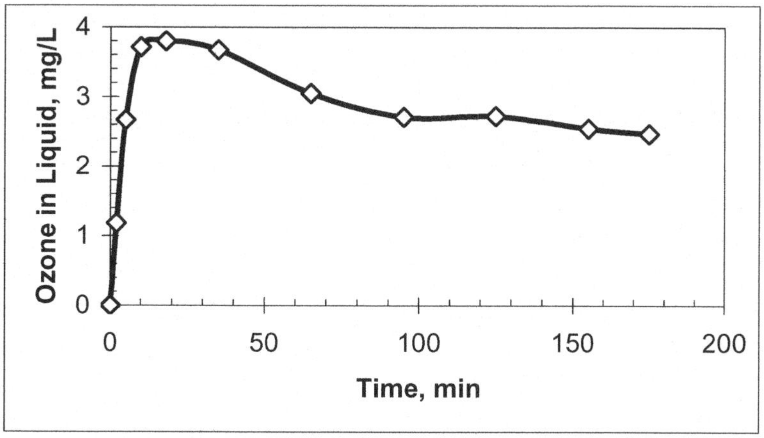

Figure 3-11. Ozone Concentration in Liquid During MTBE Ozonation.

3.2 Fenton Treatment

3.2.1 Fenton Treatment of DCDE

It is well known that Fenton reactions using ferrous or ferric salts are more effective in oxidation of organic compounds under acidic conditions. Therefore, all the experiments in this study were performed at pH values between 2.5 and 3.0. However, it should be noted that wastewaters should be thoroughly evaluated before application of the Fenton process to assure the pH decrease will not result in release of a toxic gas, e.g., hydrogen sulfide or hydrogen cyanide, creating a threat to human safety.

Several preliminary experiments were conducted to determine the effect of the concentrations of H_2O_2 and iron (as ferric sulfate) per mole of initial DCDE. The focus was to determine the molar ratio, Fe(III)/H_2O_2/DCDE, that will provide the highest amount of DCDE oxidation. The initial concentration of Fe(III) was adjusted to 1.4, 5.6 and 28 mM, while concentration of H_2O_2 was kept as 28 mM. A literature review indicates that stepwise H_2O_2 addition could be more effective than applying a large initial dose (Bowers, et. al. 1989; Mohanty and Wei, 1993). This could be due to scavenging of •OH by H_2O_2, as was also observed by Liao and Gurol (1995) for the oxidation process involving H_2O_2 and UV radiation. Stepwise addition keeps the H_2O_2 concentration at relatively low levels, reducing the detrimental effect of radical scavenging.

The results in terms of DCDE removal as a function of Fe(III) dosages are shown in Table 3-1 for three different Fe(III) dosages and for the constant H_2O_2 concentration of 28 mM. Here, the chemicals were added in four equal steps with 15-minute intervals. Increasing the concentration of iron increased the DCDE removal, as would be expected. The highest concentration of Fe(III) at 28 mM resulted in the highest removal of DCDE, yet produced the largest amount of sludge as ferric precipitates and required larger amounts of NaOH for pH restoration compared to the other ratios. However, it also resulted in the lowest residual H_2O_2 after the treatment and thus demonstrated more efficient use of H_2O_2.

Increasing the hydrogen peroxide concentration also increased the amount of DCDE oxidized. For example, 95% removal of DCDE was obtained when 35 mM Fe(III) was used with 35 mM of hydrogen peroxide. More extensive information on the effect of reagent dosages was provided earlier by Ertas and Gurol (2002) on oxidation of diethylene glycol.

Table 3-1: Fenton Treatment of DCDE for Various Reagent Dosages[1].

Fe(III)/ H_2O_2, mM/mM	28 /28 **(1/1)**	5.6 /28 **(1/5)**	1.4 /28 **(1/20)**
Fe(III)/ H_2O_2/$DCDE_o$, mM/mM/mM	19/19/1	3.5/18/1	0.9/17/1
$DCDE_o$, mM	1.45	1.58	1.63
DCDE removal, %	78	41	25
SVI[1], mL/L	390	80	16
NaOH requirement (mM) (to restore pH to 7.0)	72.8	14.5	1.9
Molar H_2O_2 consumed per mole of DCDE oxidized	24.4	40	62.5
Molar Fe(III) dose per mole of DCDE oxidized	25	8.55	3.25

[1]Results obtained after 1 hr total contact time.

In this study, oxidation of DCDE by ferric chloride and peroxide mixture was also investigated under the same conditions. Addition of 28 mM of H_2O_2 and 28 mM of $FeCl_3$ (ratio of 1/1) in four equal steps resulted in 62% removal of DCDE, compared to 78% DCDE removal obtained with $Fe_2(SO_4)_3$ under the same conditions (Table 3-2). The observation of more efficient use of iron as ferrous sulfate was also corroborated by Kiwi et al. (2000), who alluded that these results were due to the scavenging of •OH radicals by Cl^- ion, and hence the availability of less •OH radical for organic chemicals.

Table 3-2. Comparison of Results with Ferric Sulfate and Ferric Chloride.

Fe(III)/H_2O_2 : 28 mM/28 mM (Ratio: 1/1)	$Fe_2(SO_4)_3$	$FeCl_3$
$DCDE_o$, mM	1.45	1.25
% DCDE Removal	78	62
Molar H_2O_2 consumed per mole of DCDE oxidized	24.4	28.2
Molar H_2O_2 dose per mole of DCDE oxidized	25	36.4
Molar Fe(III) dose per mole of DCDE oxidized	25	36.4

3.2.2 Fenton Treatment of MTBE

As in the case of DCDE, ferric (Fe(III)) sulfate and hydrogen peroxide were added to the solutions of MTBE that contained about 200 mg/L of initial MTBE. Four different ratios of Fe(III)/ H_2O_2 were tested, keeping the H_2O_2 concentration at 28 mM: 1/20 (1.4 mM/28 mM), 1/11 (2.54 mM/28 mM), 1/5 (5.6 mM/28 mM), and 1/1 (28 mM/28 mM as before for DCDE, the chemicals were added in four steps with 15 minute intervals.

As in the case of DCDE, the maximum MTBE removal was obtained by using 1/1 ratio (28 mM/28 mM as Fe(III)/ H_2O_2). However, here MTBE was removed by only 46% (Figure 3-12) as opposed to 78% observed for DCDE. The major difference in the two systems is that the initial molar concentrations of DCDE and MTBE are respectively 1.45 mM and 2.10 mM, although the mass concentrations are equivalent at 200 mg/L. In fact, about the same number of moles of DCDE and MTBE were removed under the same treatment conditions, as shown in Figure 3-13. This indicates the formation and consumption of similar number of hydroxyl radical in both systems.

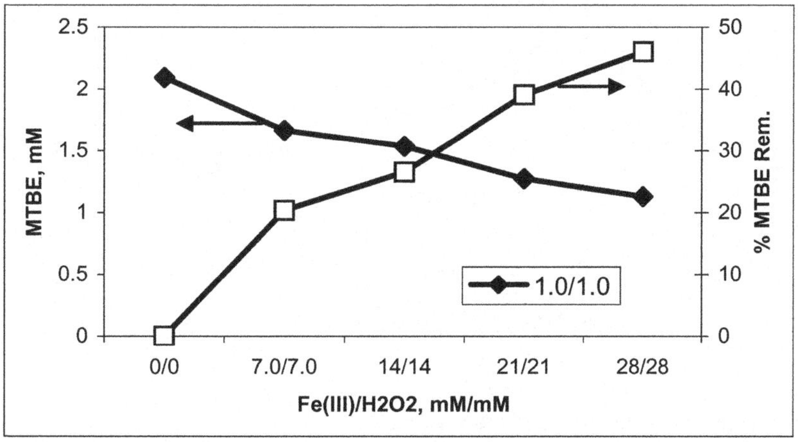

Figure 3-12. MTBE Removal by Stepwise Additions of Fe(III)/H_2O_2.

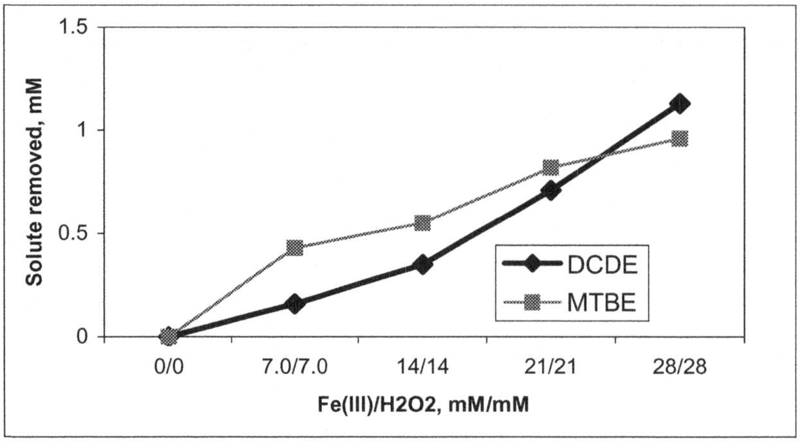

Figure 3-13. Comparison of DCDE and MTBE Removal by Stepwise Addition of Fe(III)/H_2O_2.

Although 1/1 ratio seems optimum for MTBE removal, the Fe(III) and H_2O_2 requirement to obtain 100% removal of 200 mg/L of MTBE is estimated to be excessively high. Therefore, the decision was made to test $FeSO_4 \cdot 7H_2O$ (Fe(II) instead of Fe(III)) in this treatment. By stepwise addition of Fe(II) and H_2O_2 at a ratio of 1:1, 92% MTBE removal was achieved with 7.6 mM of Fe(II) and H_2O_2 by stepwise additions in the presence of 5 mM HCO_3^- (Figure 3-14).

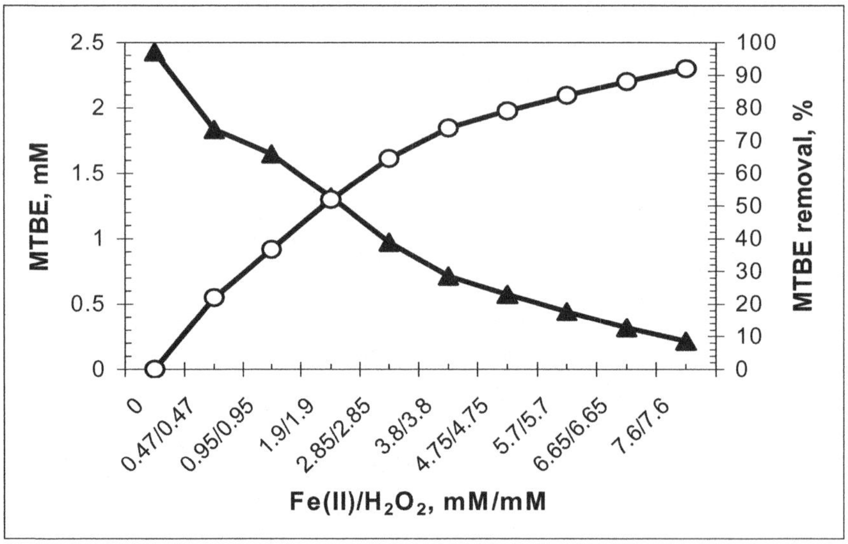

Figure 3-14. MTBE Removal with Stepwise Fe(II)/H_2O_2 Additions.

Under these conditions the reaction was completed within 60 minutes, while the pH remained at 3. In addition, all H_2O_2 reacted during the reaction. Using ferrous instead of ferric doubled the MTBE removal efficiency and still required considerably less reagents per mole of MTBE (3.3 moles of ferrous and peroxide per mole of MTBE versus 12.3 moles of ferric and peroxide per mole of MTBE).

Because of the high volatility of MTBE, its removal during Fenton treatment by volatilization was a concern. This was addressed by repeating the experiments in the absence of hydrogen peroxide. The solution containing ferrous sulfate and MTBE was mixed at acidic pH and then filtered at neutral pH to remove the iron hydroxide precipitate. Hence the removal presented in Figure 3-15 represents stripping, adsorption on precipitate and filter and volatilization during sample handling. The results showed that 40% of the initial MTBE was removed during this treatment. However, total removal of MTBE in the presence of hydrogen peroxide was 92%, indicating that significant amount of MTBE was, in fact, oxidized in the process.

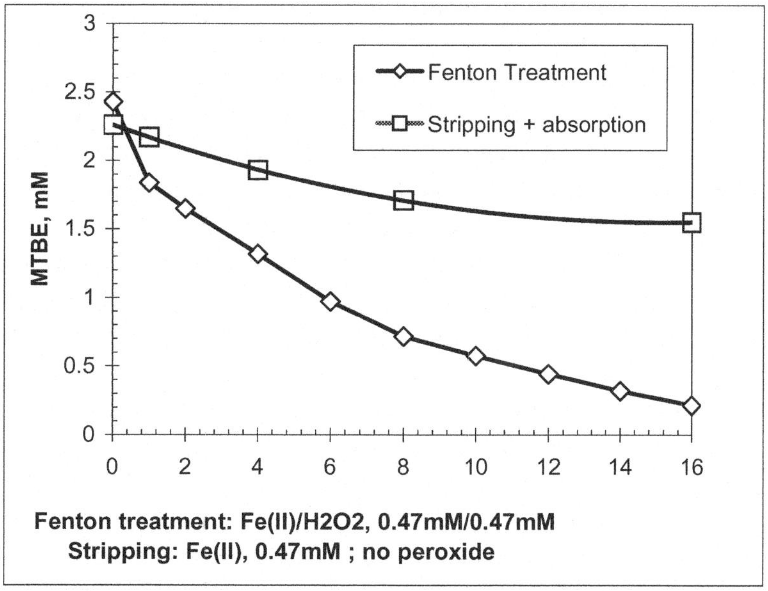

Figure 3-15. MTBE Removal During Fenton Treatment by Chemical Oxidation and Physical Removal.

3.3 UV/H_2O_2 Treatment

3.3.1 UV/H_2O_2 Treatment of DCDE

First, the photolysis effect of UV radiation on DCDE was evaluated by performing the experiments using UV light only. In these experiments, DCDE solutions were prepared with Milli-Q water at a DCDE concentration of 200 mg/L with no pH adjustment and without any carbonate. The concentration of DCDE was monitored with time of exposure. The results indicate significant reduction in DCDE concentration during the irradiation time. The DCDE concentration was reduced by 25% after 60 minutes and by 55% after 120 minutes (Figure 3-16). Monitoring of the TOC during the experiment revealed that there was no significant change in the carbon content of the solution. The pH dropped from 6.0 to about 4.0, and an increase of 4.3°C in the temperature was observed due to the heating effect of the UV lamps.

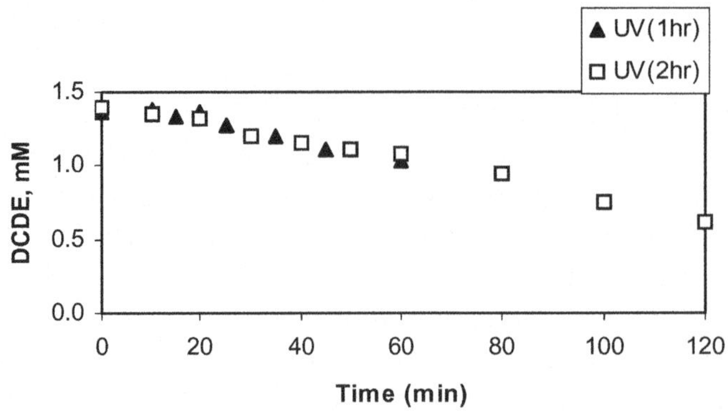

Figure 3.16. Removal of DCDE by UV Photolysis.

Several experiments were conducted to determine the effect of H_2O_2 concentration in the presence of UV light for DCDE oxidation under the same experimental condition by using non-buffered DCDE solutions of 200 mg/L. Hydrogen peroxide was added at the beginning of the experiments at DCDE/H_2O_2 molar ratios of 1.2, 0.75 and 0.65 (40, 65, 75 mg/L H_2O_2 respectively). Compared to the direct photolysis rate presented in Figure 3-16, very fast reaction rates were observed with UV/H_2O_2, as shown in Figure 3-17. This shows that DCDE reacted with hydroxyl radicals faster than it photolyzed. In Figure 3-18**,** the decrease in peroxide concentration as a function of time shows that peroxide photolyzed efficiently under these experimental conditions. Increasing the concentration of H_2O_2 from 40 to 65 increased the rate of DCDE oxidation, but further increase of the concentration to 75 mg/L did not improve the rate, indicating that 65 mg/L was the optimal H_2O_2 concentration under the experimental conditions.

The TOC removal was still negligible with H_2O_2/UV treatment. The pH drop was faster with H_2O_2/UV compared to direct photolysis as expected due to faster formation of carboxylic acids. The temperature rise was the same as in photolysis.

Additional experiments were run where hydrogen peroxide was added in a stepwise fashion in order to attain a slower reaction rate. For this procedure, 0.97 mM of H_2O_2 was added to the solution every five minutes. The lamps were turned off before each peroxide dosing. The largest drop in DCDE concentration occurred initially, and the rate was observed to be directly proportional to the H_2O_2 concentration. This stepwise procedure allowed us to oxidize DCDE by 25%, 50%, 75%, and 100% in a reasonable irradiation time. These results are presented in Figure 3-19.

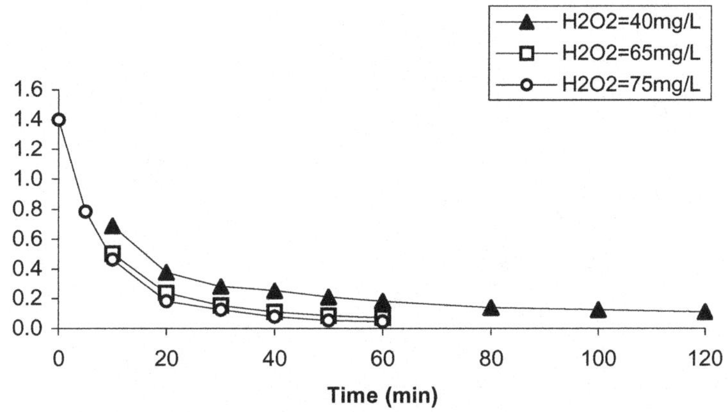

Figure 3-17. Removal of DCDE by UV/H_2O_2.

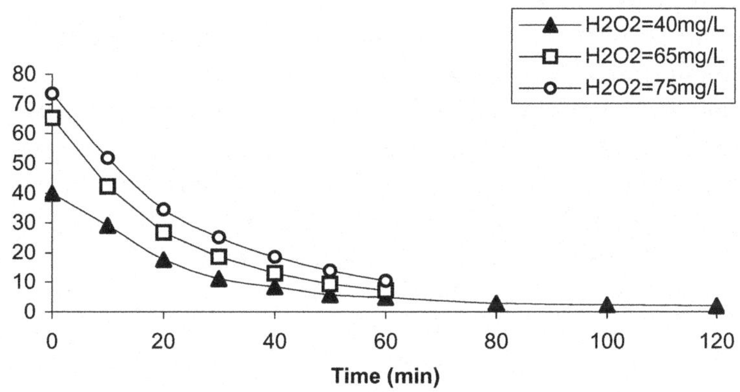

Figure 3-18. H_2O_2 Remaining after DCDE Removal.

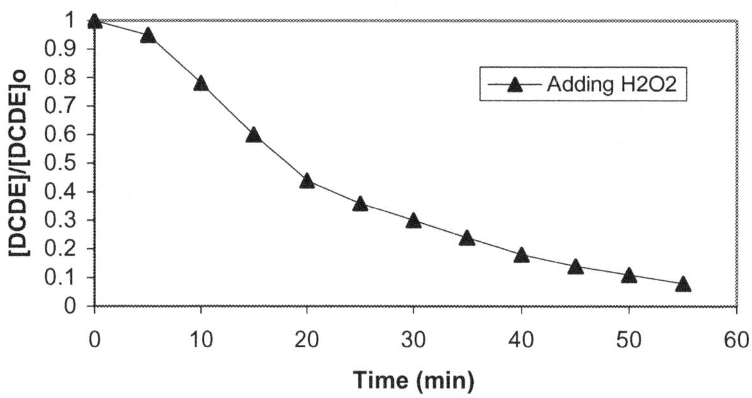

Figure 3-19. DCDE Removal by UV/H_2O_2 by Stepwise H_2O_2 Addition.

3.3.2 UV/H_2O_2 Treatment of MTBE

The UV/H_2O_2 oxidation of MTBE was conducted in the same experimental setup used for DCDE. Three liters of solution were used in each test. The solution was mixed by circulating the vessel content with a peristaltic pump. The initial MTBE concentration was 200 mg/L, initial pH was 8.4; bicarbonate concentration was 5 mM, and the initial temperature was 20°C. H_2O_2 was added stepwise as 0.97 mM of H_2O_2 every 5 minutes. This stepwise addition allowed for controlled oxidation of MTBE. Removal of MTBE by 25%, 50%, 75%, and 100% in reasonable irradiation time was achieved. The results in terms of MTBE removal are presented in Figure 3-20.

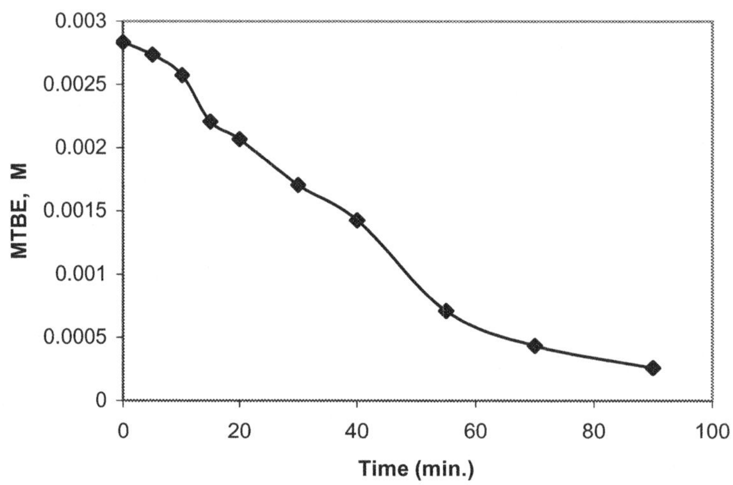

Figure 3-20. MTBE Removal by UV/H_2O_2 by Stepwise H_2O_2 Addition.

3.4 Oxidation By-Products of DCDE

Oxidation products of DCDE identified for ozonation, Fenton treatment by FeCl3 and by $Fe_2(SO_4)_3$, and UV/H_2O_2 treatment are presented in Figures 3-21, 3-22, 3-23, and 3-24, respectively.

In Tables 3-3 and 3-4, the major aldehydes and acids formed during the three different treatment methods, i.e., ozonation, Fenton treatment with both $FeCl_3$ and $Fe_2(SO_4)_3$, and UV/H_2O_2, are compared for two different levels of DCDE removal. The results are presented as percentage of mM organic carbon of the oxidation product per mM organic carbon of the initial DCDE.

Oxidation of DCDE by ozone produced formic, acetic, pyruvic, oxalic, and gyloxalic acids as the major carboxylic acids, and formaldehyde and glycolaldehyde as the major aldehyde products. When 85% of the DCDE was oxidized by ozone, about 27% of organic carbon in initial DCDE resulted in the identified aldehydes and acids. At that point, about 38% of DCDE went through dechlorination.

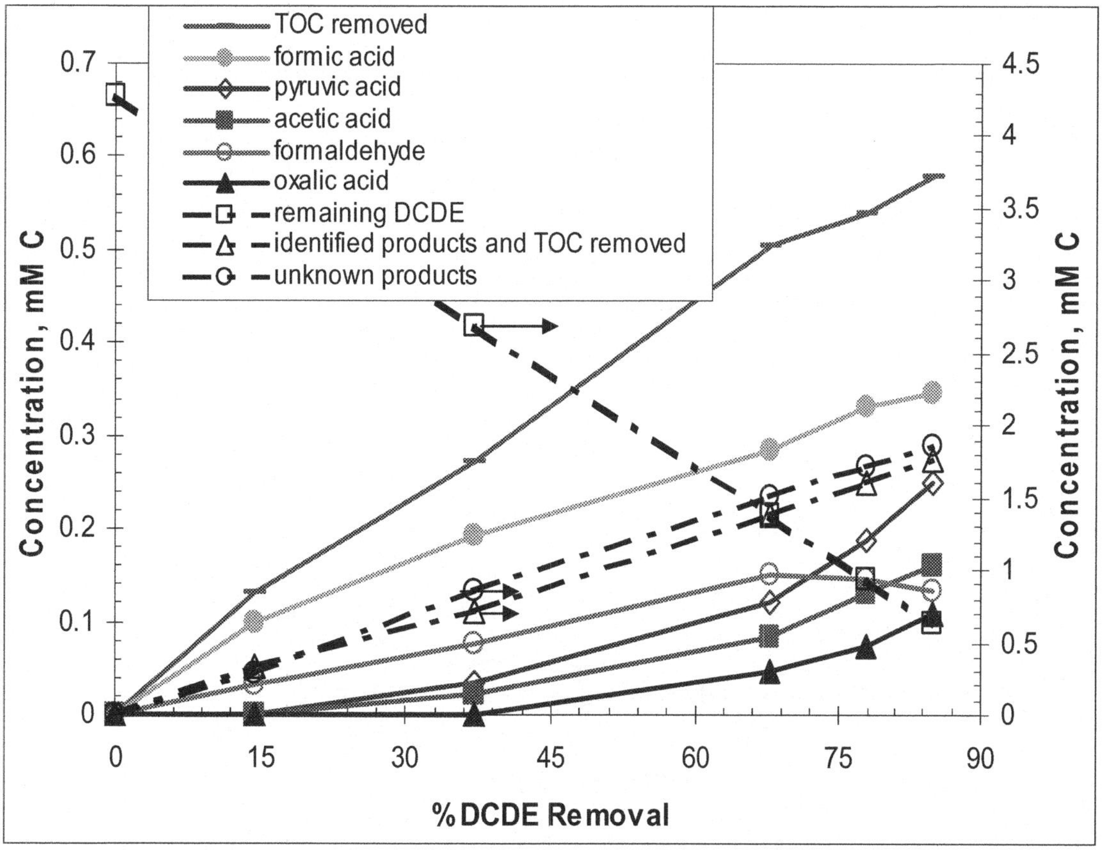

Figure 3-21. Oxidation By-Products of DCDE by Ozone. (Initial DCDE: 1.07 mM)

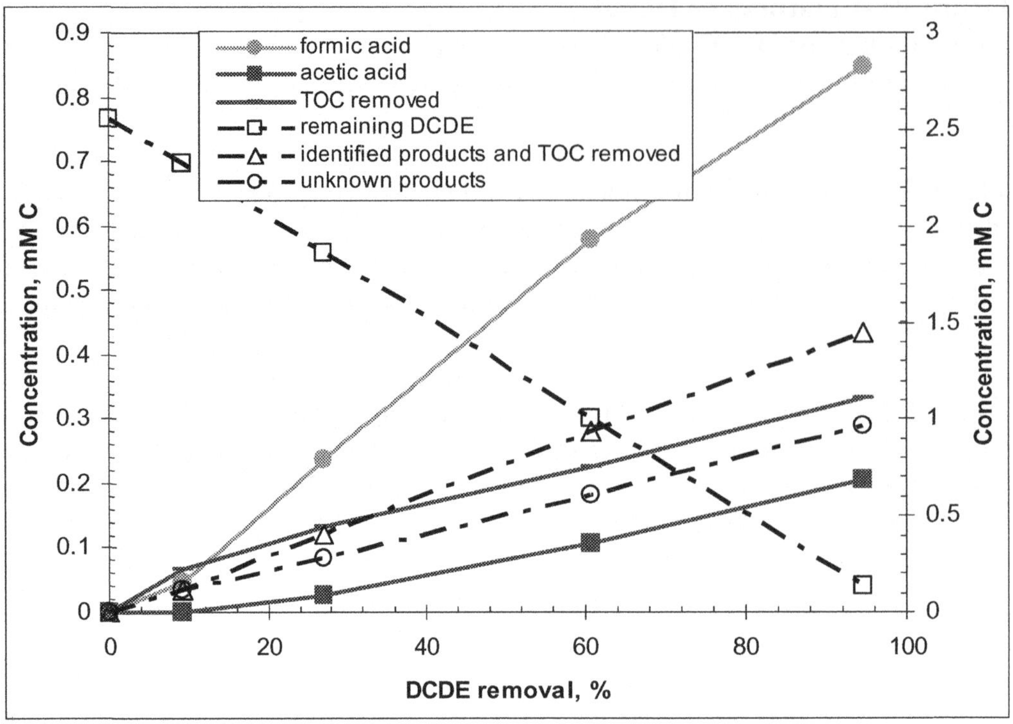

Figure 3-22. Oxidation By-Products of DCDE of Fenton Treatment by $FeCl_3$. (Initial DCDE:0.64 mM)

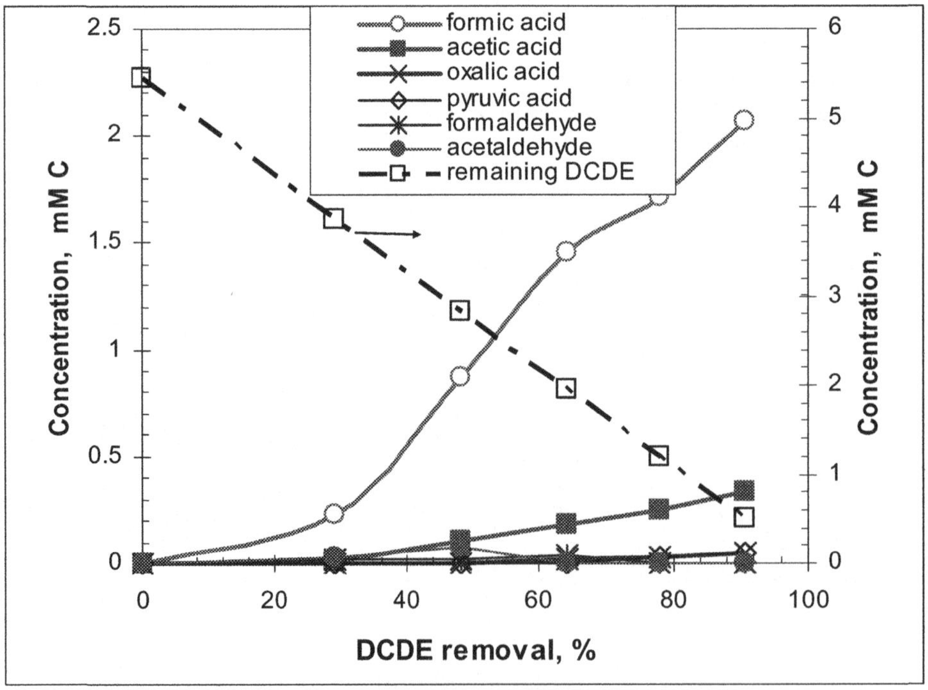

Figure 3-23. Oxidation By-Products of DCDE of Fenton Treatment by $Fe_2(SO_4)_3$. (Initial DCDE: 1.36 mM)

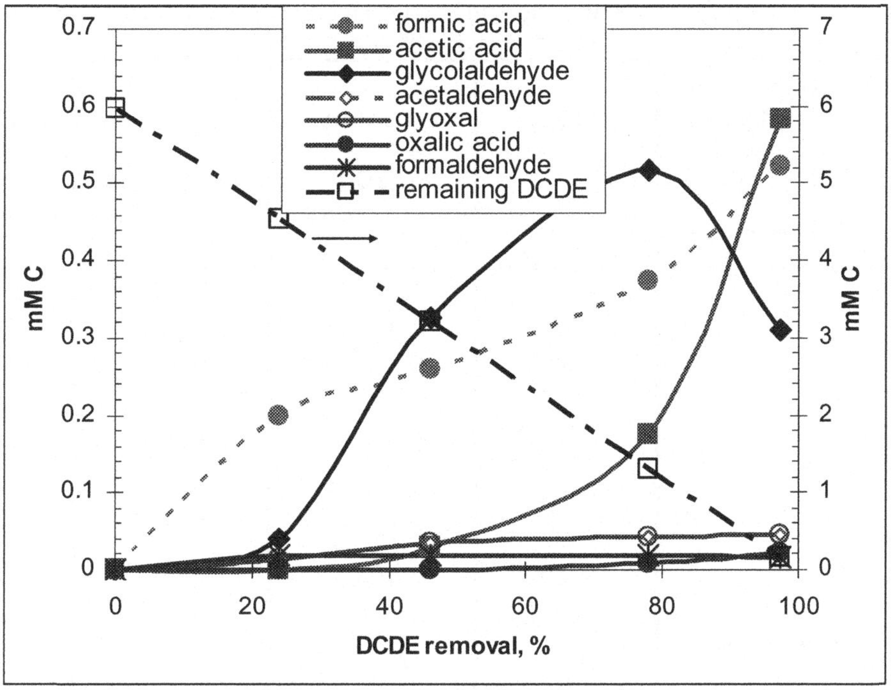

Figure 3-24. Oxidation By-Products of DCDE by UV/H_2O_2.

The results show no significant differences for the types and the concentrations of the aldehydes and acids produced by Fenton treatment using either $FeCl_3$ or $Fe_2(SO_4)_3$ as the iron salt. Furthermore, during Fenton treatment of DCDE with both iron salts, formic acid was produced as the primary reaction product; small amounts of acetic acid were also formed during the reaction. When about 91–95% of the DCDE was oxidized by Fenton reagent, about 45% of the organic carbon in the initial DCDE resulted in the identified aldehydes and acids. At that point, about 54% of the DCDE dechlorination was achieved with both $FeCl_3$ or $Fe_2(SO_4)_3$.

Oxidation of DCDE by H_2O_2/UV produced mainly formic and acetic acids as the major carboxylic acids and glycolaldehyde as the major aldehyde product. When 97% of the DCDE was oxidized by H_2O_2/UV, about 27% of organic carbon in the initial DCDE resulted in the identified aldehydes and acids. At that point, about 54% of the DCDE went through dechlorination.

Table 3-3. Comparison of Intermediate Products for All Oxidation Methods. (DCDE Removal: 61%–78%)

Intermediate Products[1]	Oxidation Method and % DCDE Removal	Ozone	Fenton ($FeCl_3$)	Fenton ($Fe_2(SO_4)_3$)	UV/H_2O_2
		78%	61%	78%	78%
Formaldehyde		3.41	0.28	0.06	0.29
Glycolaldehyde		1.70	0.02	0.02	8.67
Glyoxal		0.77	0	0	0.71
Acetaldehyde		1.23	0.28	0	3.28
Formic acid		7.79	22.68	31.38	6.26
Oxalic acid		1.76	0	0.49	0.11
Acetic acid		3.07	4.22	4.62	2.92
Pyruvic acid		4.39	0.57	0.59	0
Glyoxalic acid		1.06	0	0	0
Total Intermediate Products		25.18	28.05	37.16	22.25
Dechlorination, %		32.6	30.7	36.1	29.7

[1] (mM C IP/mM C DCDEo)x100;

Table 3-4. Comparison of Intermediate Products for All Oxidation Methods. (DCDE Removal: 85%–97%)

Intermediate Products[1]	Oxidation Method and % DCDE Removal	Ozone	Fenton ($FeCl_3$)	Fenton ($Fe_2(SO_4)_3$)	UV/H_2O_2
		85%	95%	91%	97%
Formaldehyde		3.12	0.64	0	0.28
Glycolaldehyde		1.37	0.08	0	5.19
Glyoxal		0.77	0	0	0.77
Acetaldehyde		0.97	0.32	0	1.55
Formic acid		8.13	33.20	37.92	8.79
Oxalic acid		2.54	1.07	0.84	0.35
Acetic acid		3.79	8.01	6.22	9.79
Pyruvic acid		5.85	0.34	0.80	0
Glyoxalic acid		1.31	0	0	0
Total Intermediate Products		27.87	43.66	45.77	26.74
Dechlorination, %		38.2	55.1	52.4	54.3

The product distribution for the three oxidation method showed that oxidation of DCDE progressed by different pathways. For example, formaldehyde formed primarily with ozone. formic acid formed primarily for Fenton reagent, and glycolaldehyde formed primarily with UV/H_2O_2.

Overall, ozonation and UV/H_2O_2 treatment produced higher percent conversions to aldehydes, whereas both Fenton treatments produced higher percent conversions to carboxylic acids for the same levels of DCDE oxidation. Yet, UV/H_2O_2 produced higher conversions glycolaldehyde and acetaldehyde than ozonation. These observations imply very different reaction mechanisms for oxidation by ozone, Fenton's reagent, and UV/H_2O_2. This, to some degree, contradicts the expected result that ozone, Fenton reagent, and UV/H_2O_2 should produce the same products and at the same conversion levels. This is because ozone was observed to react with DCDE primarily through hydroxyl radical in this study, and Fenton's reagent and UV/H_2O_2 are also expected to react through formation of hydroxyl radical. On the other hand, Bossman et. al., (1998) recently observed completely different oxidation products of 2,4-xylidine by UV/H_2O_2 and Fenton reagent, and concluded that Fenton oxidation involves an electron transfer mechanism rather than formation of hydroxyl radical, supporting our observations in this study.

The concentration of chloride ion produced as a result of DCDE oxidation is presented in Figure 3-25 for three oxidation methods, using both iron salts for Fenton reagent.

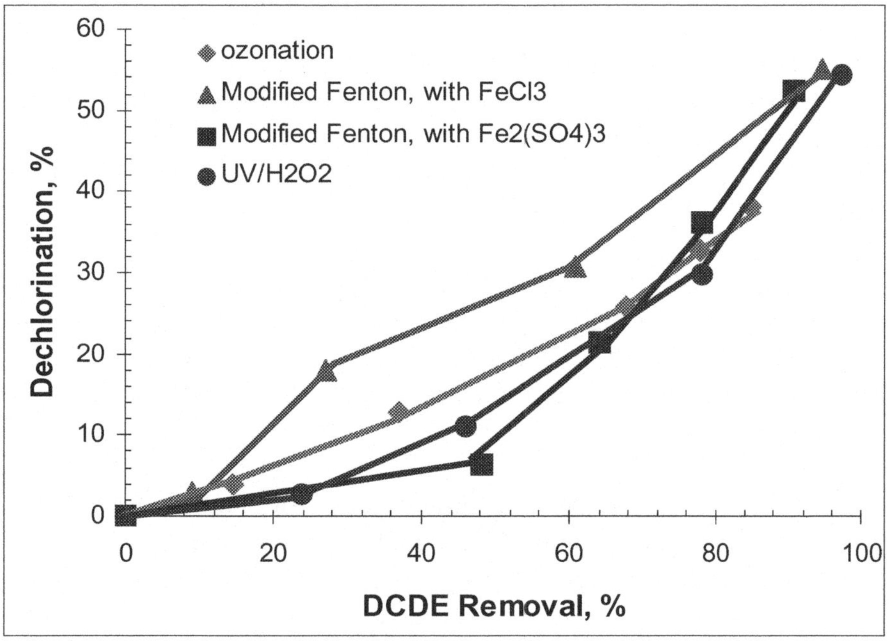

Figure 3-25. Dechlorination of DCDE During Its Oxidation.

The results indicate no significant differences among the treatment methods. However, only about 55% of the original chloride was recovered as chloride ion when 85%–97% of the DCDE was oxidized, indicating that large amounts of the chloride was probably still present in unidentified chlorinated products. The TOC removed, i.e., organic carbon mineralized during DCDE oxidation by H_2O_2/UV, was less than 10% of the initial organic carbon present, a result which agreed with both ozonation and Fenton treatment.

Additional product information was sought to provide a more complete material balance. The analytical protocol was modified in order to detect and quantify any chlorinated organics that may have escaped detection. Attention was focused on chlorinated compounds because they are often more recalcitrant to biological degradation and show higher toxicity then their non-chlorinated analogs.

First, extraction procedures and analytical techniques were modified to look for non-polar organics, trace aldehydes, and trace organic acids. Secondly, all previous data were reviewed and unidentified compounds were investigated. DCDE (200 mg/L) was oxidized by ozonation. Samples were taken during the course of oxidation and collected in 40 mL volatile organic analysis (VOA) vials. The samples were split. A 20 mL sample was extracted with 2 mL of hexane. The hexane was decanted off and analyzed by gas chromatography/mass spectrometry (GC/MS) for neutral compounds. The remaining aqueous phase was acidified ($pH < 2$) and extracted with MTBE. The MTBE fraction was methylated using trimethylsilyl diazomethane (TMSD). Methylation was done to aid in the detection of organic acids such chloro acetic acid. The methylated fraction was analyzed by MS. A separate 20 mL sample was derivatized using a second method, PFBHA, which is more sensitive to aldehydes. A third fraction was diluted (1:5) and analyzed by ion chromatography.

Analysis of the neutral fraction showed one major chlorinated product. This product was also formed during UV/H_2O_2 and Fenton oxidation. This peak was identified as (2-chloroethoxy) chloroacetaldehyde. Although the fragments show only a single chlorine atom, the structure proposed has two chlorines. The second chlorinated peak detected was *bis*(chloroethoxy)ethane, a contaminant present in the DCDE.

The methylated samples showed an additional peak that co-eluted with the (2-chloroethoxy)chloroacetadehyde peak. This peak was tentatively identified as (2-Chloroethoxy) acetic acid methyl ester. The un-derivatized product was (2-chloroethoxy) acetic acid. Aldehydes were analyzed for using the same PFBHA derivatization method. The results are presented in Table 3-5.

It should be noted that these chlorinated products were not quantified primarily because of lack of standard solutions for many of them. It is also possible that additional chlorinated reaction by-products may exist in solutions and can perhaps be identified by different techniques than the one used in this study.

Table 3.5. DCDE Oxidation Products Identified by Mass Spectrometry and IC.

Product		UV/H_2O_2	Fenton	O_3
(Chloroethoxy)chloroacetaldehye	$ClCH_2CH_2OC(O)CH_2Cl$	X	X	X
(2-Chloroethoxy) Acetic acid	$ClCH_2CH_2O$- $CH_2C(O)OHCl$	ND	ND	X
Chloroglycoaldehyde	$ClC(O)CH_2OH$	X	ND	ND
Chloroethoxy ethanol	$ClCH_2CH_2OCH(OH)CH_3$	X	ND	ND
Bis(chloroethoxy) ethane	$ClCH_2CH_2OCH_2CH_2O$-CH_2CH_2Cl	X	X	X
$C_7H_9O_3Cl_2$ ret time=36.1 min	$C_7H_9O_3Cl_2$	X	ND	ND
$C_7H_9O_3Cl_2$ ret time = 36 2 min	$C_7H_9O_3Cl_2$	X	ND	ND
Formaldehyde	HC(O)H	X	X	X
Acetone	$H_3CC(O)CH_3$	ND	X	X
Acetaldehyde	$H_3CC(O)H$	X	X	X
Glyoxal	HC(O)C(O)H	X	ND	X
Glycolaldehyde	$HOCH_2C(O)H$	X	ND	X
Formic Acid	HOC(O)H	X	X	X
Oxalic Acid	HOC(O)C(O)OH	X	ND	X
Acetic Acid	$H_3CC(O)OH$	X	X	X
Chloroacetic acid	$ClCH_2C(O)OH$	ND	ND	X
Pyruvic Acid	$H_3CC(O)C(O)OH$	X	X	X
Glyoxalic acid	HOC(O)C(O)H	X	ND	X

X = compound detected during oxidation. ND = not detected.

3.5 Oxidation By-Products of MTBE

Oxidation products of MTBE identified during ozonation are presented in Figures 3-26 and 3-27 in units of mg/L versus ozonation time, and mM organic carbon versus percent MTBE removal, respectively for 110 mg/L of initial MTBE concentration.

TBA, tertiary butyl formate (TBF), and acetone were detected as the major intermediate products upon ozonation of 110 mg/l of MTBE. The only aldehyde detected was formaldehyde. Formic acid was also detected as an acidic product. But both formaldehyde and formic acid concentrations were an order of magnitude lower than the concentrations of TBA, TBF, and acetone. Very low amounts of pyruvic (3.8×10^{-4} mM as C) and acetic (1×10^{-3} mM as C) acids were also detected. Total intermediate product concentration obtained at the end of 180 minutes of reaction time was 1.7 mM as C. In addition, about 3.4 mM of organic carbon was removed during this period. Hence, by-products, plus TOC removed, plus remaining MTBE can account for 95% of the organic carbon contributed by the initial MTBE concentration of 6.2 mM as C. It should be mentioned that a large fraction of TOC removed was due to volatilization of MTBE, as shown in Figure 3-9.

The intermediate products of Fenton treatment of MTBE are presented in Figures 3-28 and 3-29 in mg/L and mM carbon as a function of percent MTBE removal.

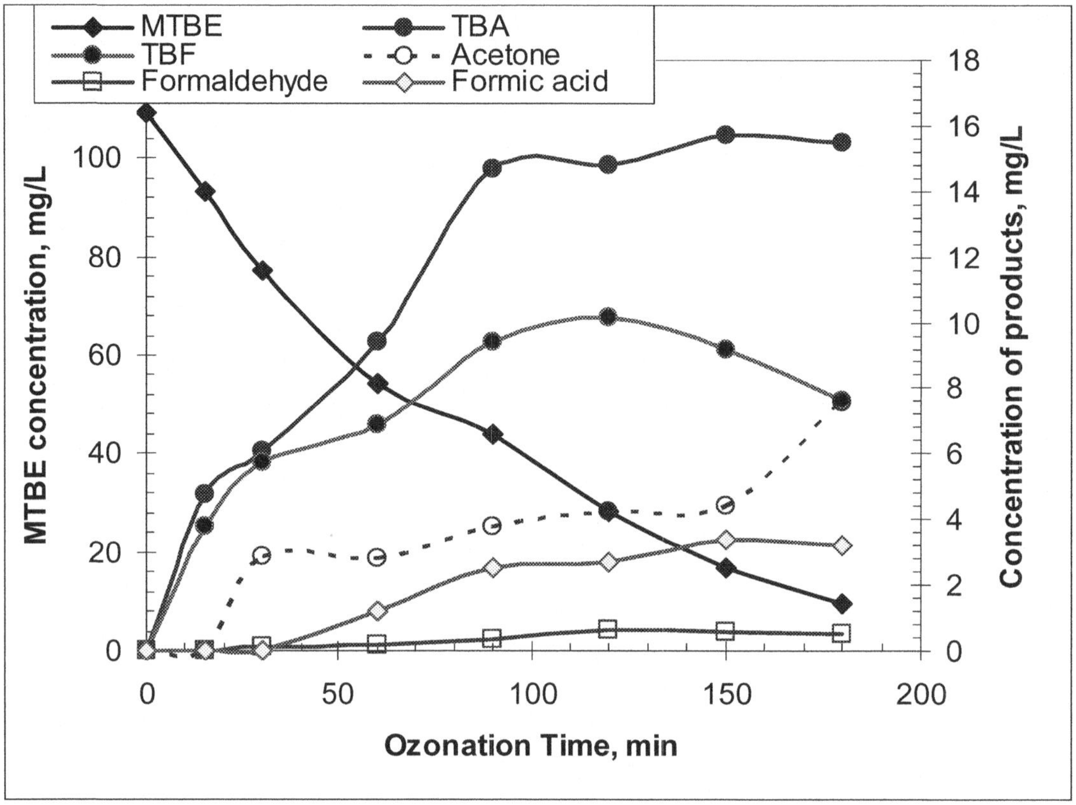

Figure 3-26. Major Intermediate Products of Ozonation of MTBE in mg/L.

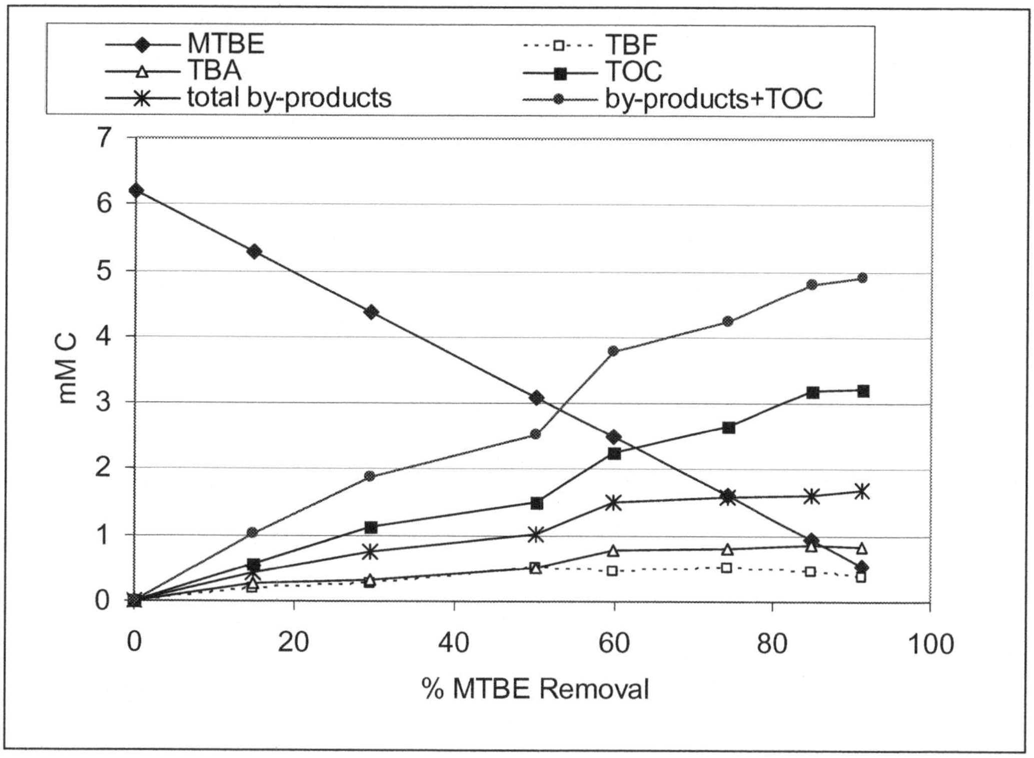

Figure 3-27. Major Intermediate Products of Ozonation of MTBE in mM C.

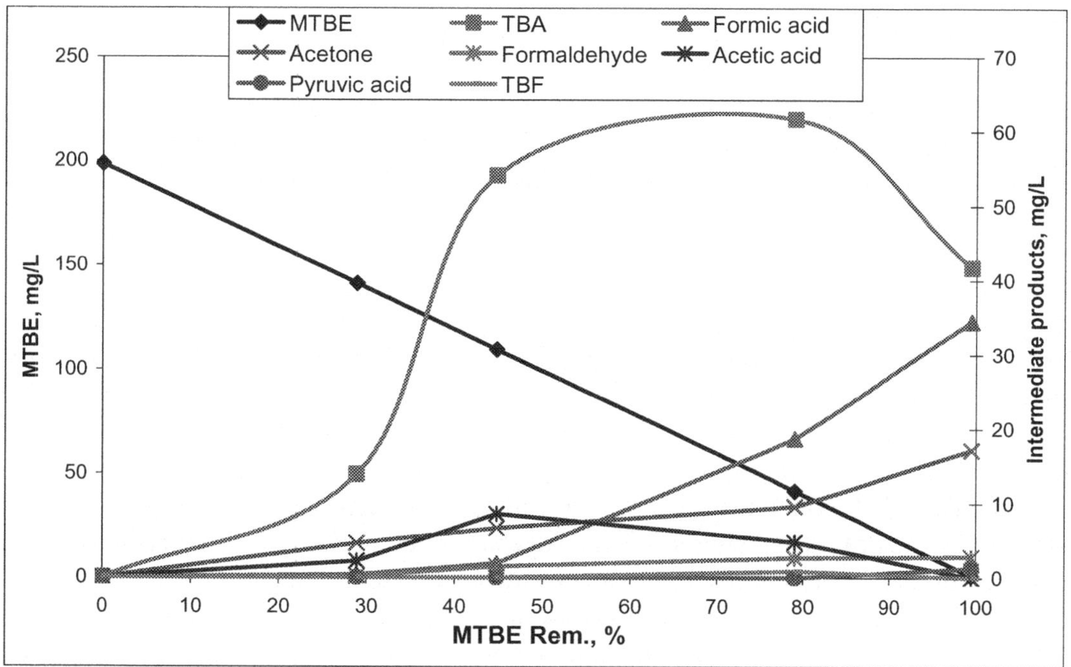

Figure 3-28. Major Intermediate Products of Fenton Treatment of MTBE in mg/L.

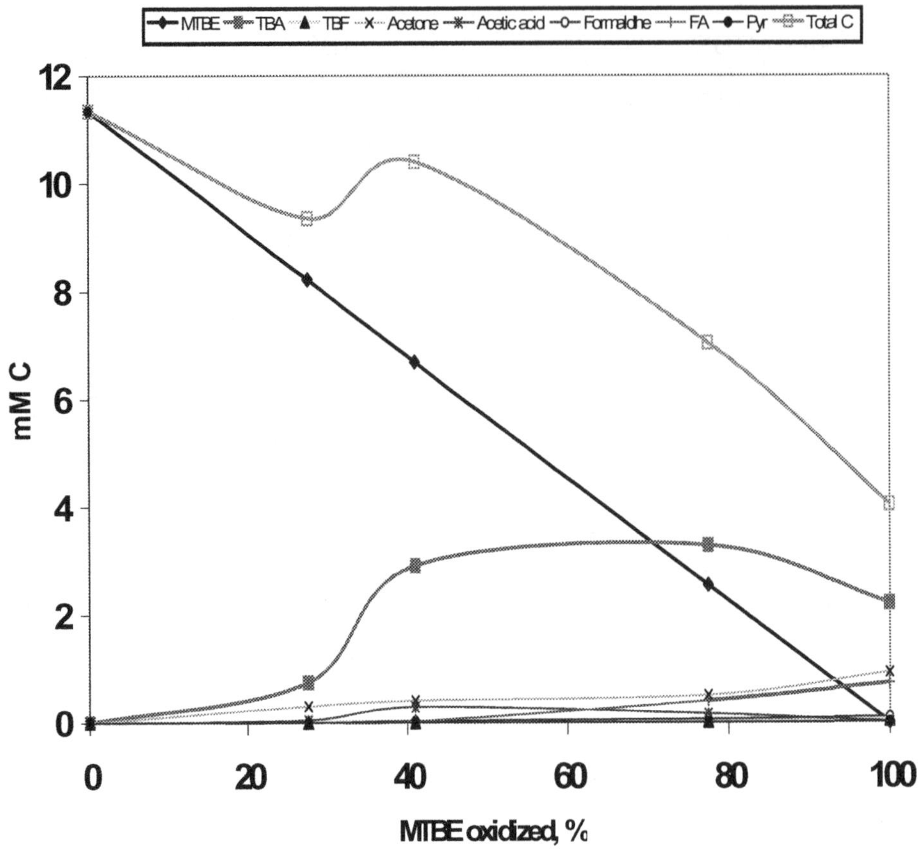

Figure 3-29. Major Intermediate Products of Fenton Treatment of MTBE in mM C.

In comparing the products of ozonation and Fenton treatment, it should be noted that the initial MTBE concentrations were not the same. Furthermore, large amounts of MTBE were stripped during ozonation without getting oxidized. These might explain the formation of low concentrations of the ozonation products.

The intermediate products of UV/H_2O_2 treatment of MTBE are presented in Figures 3-30 and 3-31 in mg/L and mM carbon as a function of percent MTBE removal. TBA, acetone, and methanol appeared very early in the reaction. Acetone production leveled off quickly, and acetic acid increased. Acetone is the logical precursor to acetic acid. Methanol was observed during UV/H_2O_2 oxidation but not during ozonation or Fenton oxidation. Hong et al. (1999) reported it as a product of MTBE from Fenton oxidation. Methanol production appears to be related to the formation of TBA. TBF concentrations increase until approximately 50% removal of MTBE. The expected product of TBF oxidation is TBA, and indeed, TBA remained relatively high until TBF concentrations declined. Only low concentrations of methyl acetate were observed. This is in agreement with Stefan et al. (2000). In Figure 3-31 the curve labeled "TOC" is the difference in the measured TOC and the initial TOC and is a measure of CO_2 evolution. The top line is a sum of the theoretical TOC from the measured products plus the change in TOC. A good mass balance was maintained until the later stages of oxidation. Compared to ozonation, the products were more completely oxidized in the UV/H_2O_2 process.

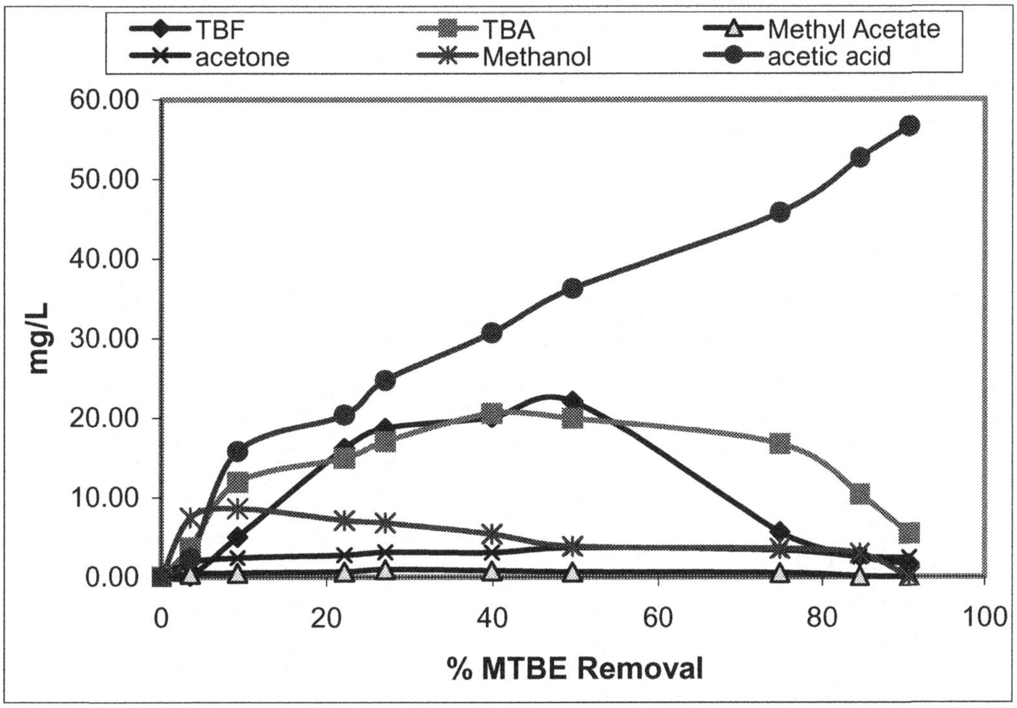

Figure 3-30. Intermediate Products of UV/H_2O_2 Treatment of MTBE in mg/L.

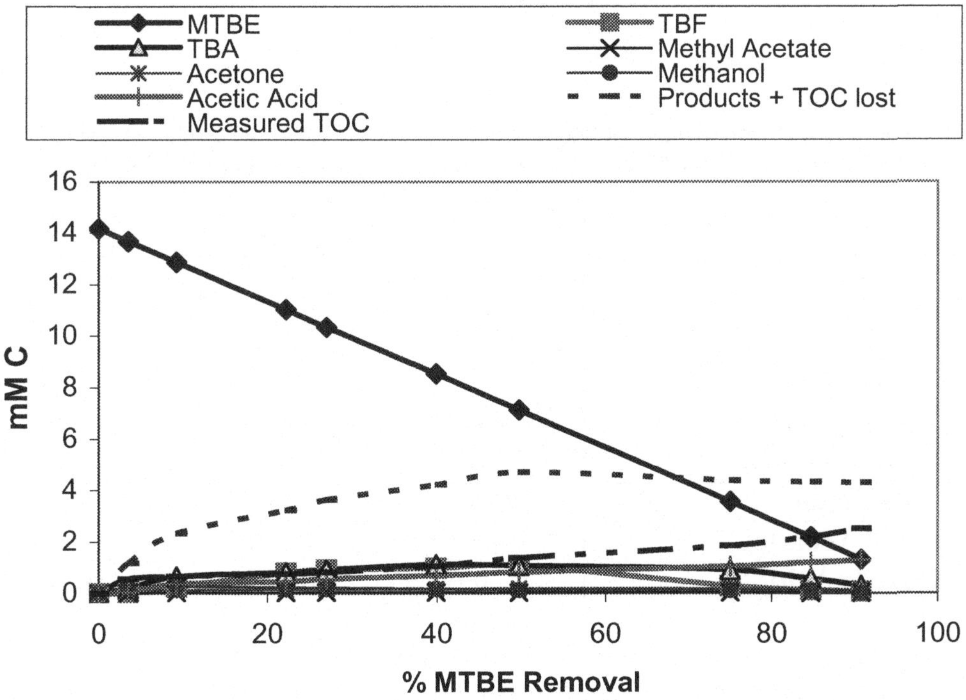

Figure 3-31. Major Intermediate Products of UV/H_2O_2 Oxidation of MTBE in mM C.

In Table 3-6, the product distributions of the three oxidation methods are compared at an MTBE removal level of 80%–90%. The intermediate products are given in mg/L. The initial concentration of MTBE in Fenton and UV/H_2O_2 processes is 11.2 mM, whereas in ozonation process, it is 6.2 mM. The product concentrations were normalized per 100 moles of MTBE oxidized and presented in Table 3-7.

Table 3-6. Comparison of Intermediate Products from MTBE for Three Oxidation Methods (MTBE Removal: 80%–90%). Product Concentrations in mg/L.

	Ozone[1]	Fenton[2]	UV/H_2O_2[3]
TBA	15.6	62	10.43
TBF	9	0.2	2.61
Acetone	4.5	0.5	2.58
Acetic Acid	0	0	52.68
Formaldehyde	0.6	0.3	0
Formic Acid	3.5	19	0
Methanol	N.M.	N.M.	2.99
Methyl Acetate	N.M.	N.M.	0.11
Pyruvic Acid	0	0.1	0

[1] $MTBE_o$: 6.2 mM; [2, 3] $MTBE_o$: 11.2 mM

Table 3.7. Comparison of Intermediate Products from MTBE for Three Oxidation Methods (MTBE Removal: 80%–90%). Normalized to Moles per 100 moles of MTBE Removed.

	Ozone	Fenton	UV/H_2O_2
TBA	19.7	43.4	7.29
TBF	8.24	0.102	1.34
Acetone	7.24	0.451	2.33
Acetic Acid	0	0	45.8
Formaldehyde	1.86	0.521	0
Formic Acid	7.09	21.6	0
Methanol	N.M.	N.M.	4.89
Methyl Acetate	N.M.	N.M.	0.081
Pyruvic Acid	0	0.0595	0

A general trend can be seen where ozonation produced more TBF, acetone, and formaldehyde than the other processes. Fenton process, on the other hand, produced more TBA and formic acid. Acetic acid was observed only in UV/H_2O_2 oxidation.

Different product distribution may indicate different oxidation pathways for the three methods. On the other hand, reactor configurations can partially explain different product distribution. The Fenton oxidation reactor is a batch-mixed reactor. Hydroxyl radicals are expected to form uniformly throughout the reactor, and those compounds that react faster with OH radicals disappear first and slower reacting compounds accumulate. The ozonation and UV/H_2O_2 reactors are semi-batch reactors, where one reactant, ozone, or UV radiation, is added continuously. Thus the dissolved ozone concentration might be somewhat higher at the bottom of the reactor near the diffuser where the bubble size is smallest and the gas concentration is highest, than near the upper regions. In the UV/H_2O_2 reactor, the incident UV intensity decreases radially from the outside to the center. All three systems were well mixed of course, but hydroxyl radicals react at near diffusion-controlled rates ($1 \times 10^9\ M^{-1}s^{-1}$). Thus it is possible that localized concentration gradients of hydroxyl radicals might be established in the semi-batch reactors.

CHAPTER 4.0

ASSESSMENT OF BIODEGRADATION FOR DCDE

4.1 Short-Term Biodegradation Tests

The tests conducted on a range of different concentrations of non-oxidized DCDE indicated that DCDE was not biodegradable by non-acclimated activated sludge culture. Microorganisms exposed for a period of more than four weeks to 0.4 mg/L of DCDE and thereby presumably acclimated to its presence were not able to degrade non-oxidized DCDE either. Only trivial exogenous respiration rates were measured (data not shown, but available in Kaludjerski, 2001), confirming the results previously reported in the *Hazardous Chemicals Desk Reference* that DCDE is nonbiodegradable in conventional biological treatment systems (Lewis, 1991).

DCDE solutions preoxidized at 25%, 50%, 75%, and 100% were subjected to short-term biodegradation experiments after diluting with mineral salt solution and distilled water at different proportions. The proportions were adjusted to result in a relatively wide range of biodegradable substrate (non-parent TOC concentration). Figure 4-1 shows typical raw data for an experiment where time is in seconds.

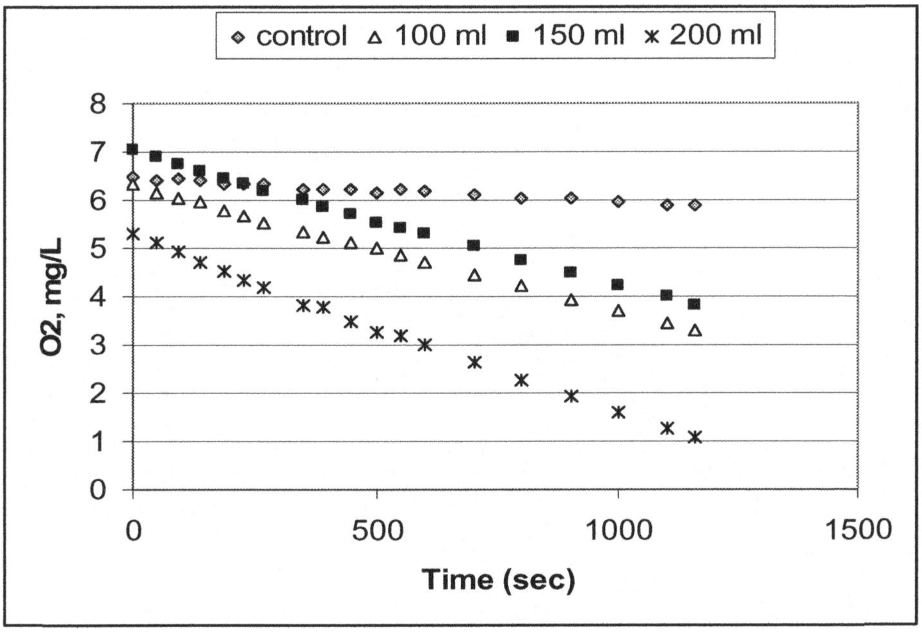

Figure 4-1. Respiration Rates of Microorganisms with Different Dilutions of DCDE After 25% of Oxidation by Ozone.

After net exogenous respiration rate for each experiment was estimated, as described in Section 2.2.4, the data were plotted as a function of non-parent TOC, i.e., the organic matter produced as oxidation by-products.

4.1.1 Ozone-Treated DCDE

In Figure 4-2 and Figure 4-3**,** the data for ozonated DCDE are presented as a function of non-parent TOC for non-acclimated and acclimated cultures, respectively. The varied parameter in the figures is the percentage of DCDE oxidation prior to the biodegradation tests. The data point for non-oxidized DCDE, i.e., zero amount of non-parent TOC, is represented by the point of origin in the figures.

The results show that an increase in the substrate concentration is reflected by a rapid increase in the net respiration rates, reaching a plateau at higher concentrations, according to the Monod equation. The net respiration rates for all levels of oxidation are significantly higher than the rate of non-oxidized DCDE, which is zero. This increase in the respiration rates is clearly due to existence of biodegradable chemicals produced as by-products of DCDE oxidation by ozone. For ozonated solutions subjected to non-acclimated culture (Figure 4-2), different ozonation levels produced similar results in terms of enhancement in the respiration rates. A slight drop in the maximum respiration rate (q_{max}) can be seen in the samples oxidized by 100%. However, acclimated culture produced significantly different results for ozonated solutions. The solution ozonated for 75% removal of DCDE produced a significantly higher respiration rate than the other solutions. Furthermore, the respiration rate dropped significantly for 100% oxidized DCDE. The decreased respiration rate for 100% oxidation can be explained by formation of higher concentrations of by-products from oxidation of DCDE that might either be toxic or less biodegradable.

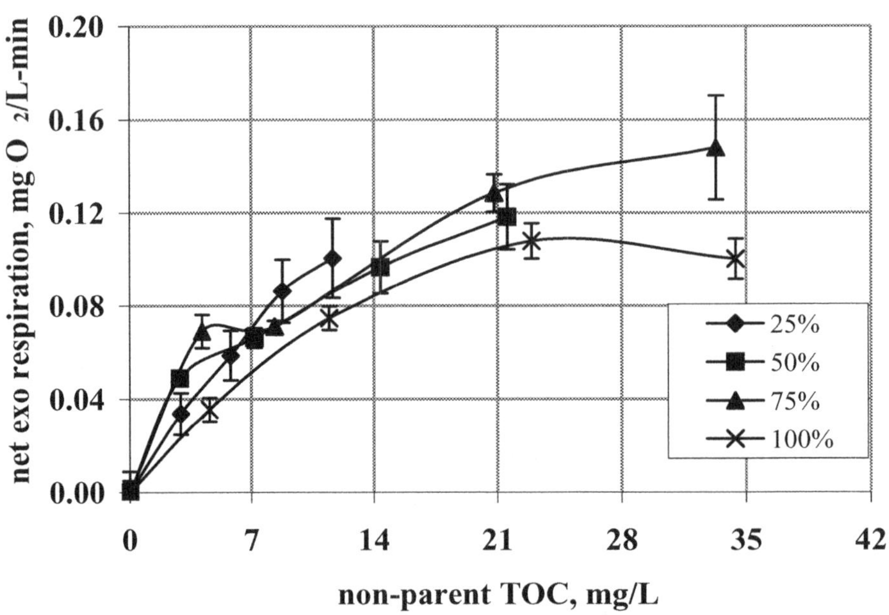

Figure 4-2. Respiration Rates of Non-Acclimated Cultures for Ozone-Treated DCDE Solutions.

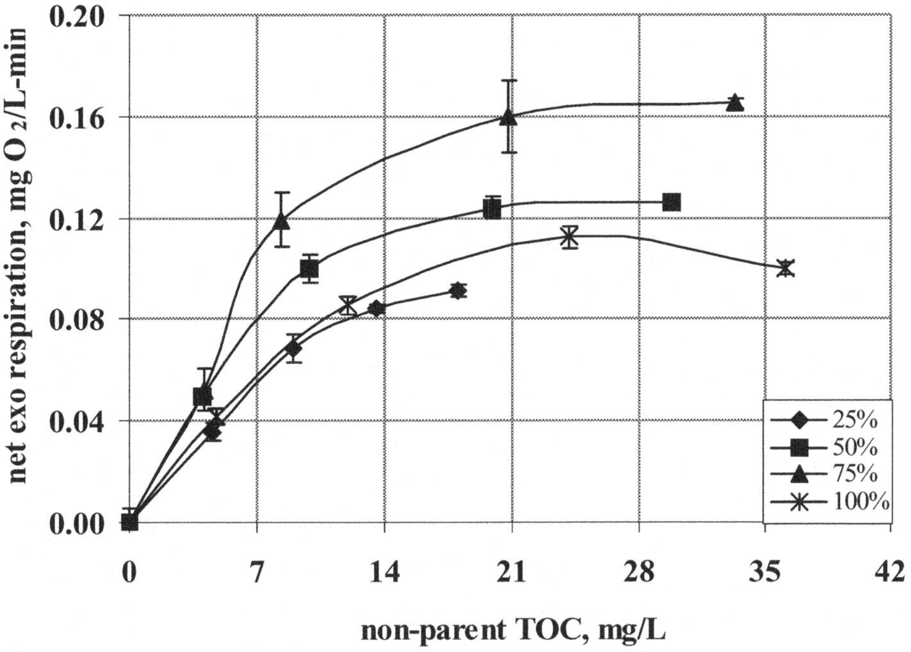

Figure 4-3. Respiration Rates of Acclimated Cultures for Ozone-Treated DCDE Solutions.

4.1.2 Fenton-Treated DCDE

The results for DCDE solutions oxidized by Fenton reagent are presented in Figures 4-4 and 4-5. The results are not very different than those obtained for ozonated solutions, i.e., the net respiration rate increased with increasing substrate concentration. For non-acclimated culture, higher oxidation rates, e.g., 75% and 100%, enhanced the respiration rate to a greater extent than did the lower oxidation rates (Figure 4-4). However, for the acclimated culture (Figure 4-5), a lower maximum respiration rate was observed for 100% oxidized solution than the others. These samples are again highly oxidized and, therefore, possibly contain higher concentrations of the by-products that are either toxic or less biodegradable. The acclimated cultures seem to be more sensitive to the oxidation by-products of DCDE. This higher sensitivity suggested for the acclimated cultures could be due to reduced resistance in a less diverse microbial culture.

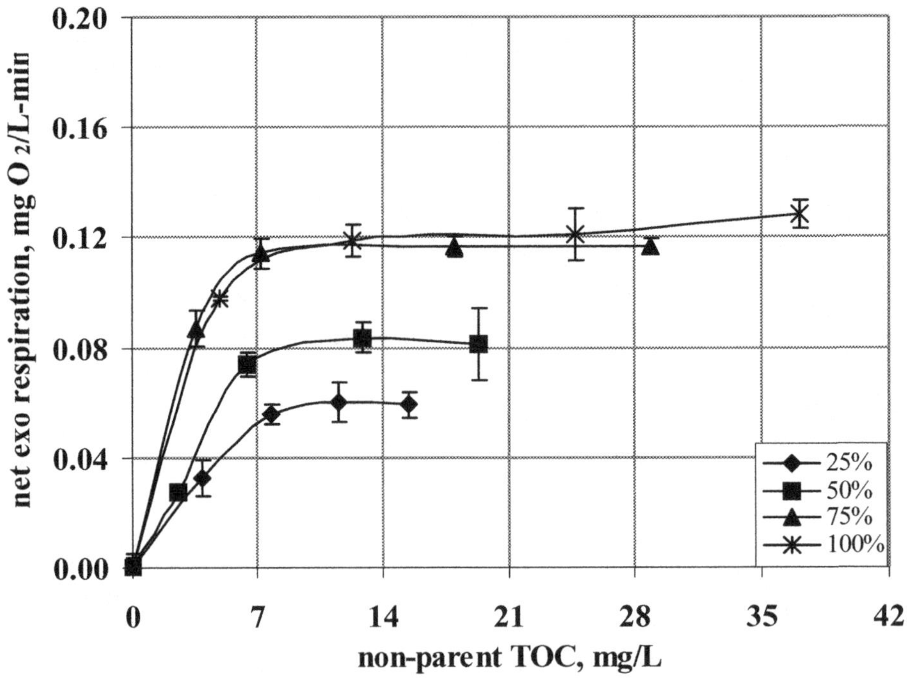

Figure 4-4. Respiration Rates of Non-Acclimated Cultures for Fenton-Treated DCDE.

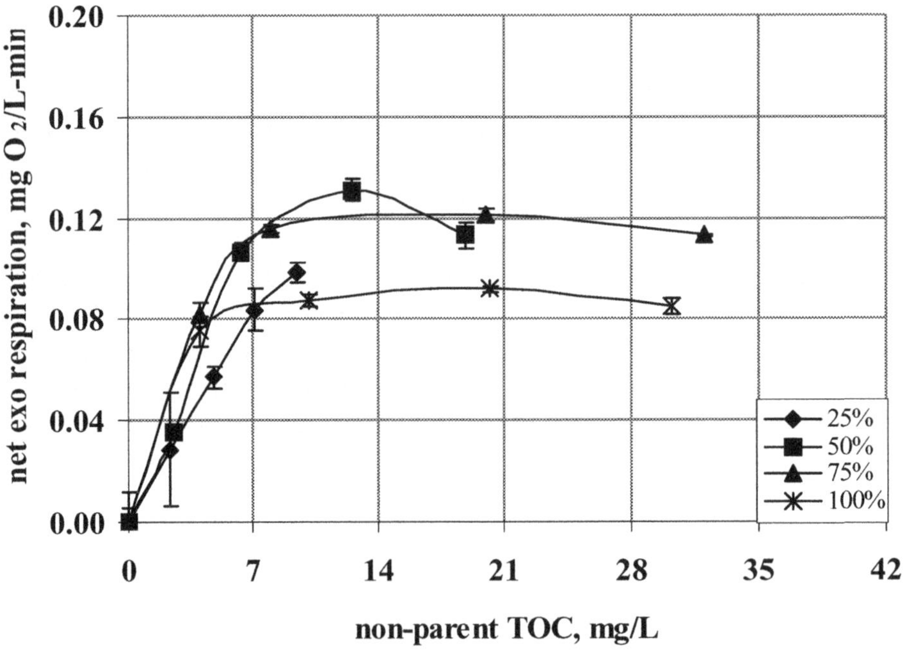

Figure 4-5. Respiration Rates of Acclimated Cultures for Fenton-Treated DCDE.

4.1.3 UV/ H_2O_2-Treated DCDE

The results of the short-term tests for UV/H_2O_2-treated DCDE are presented in Figures 4-6 and 4-7.

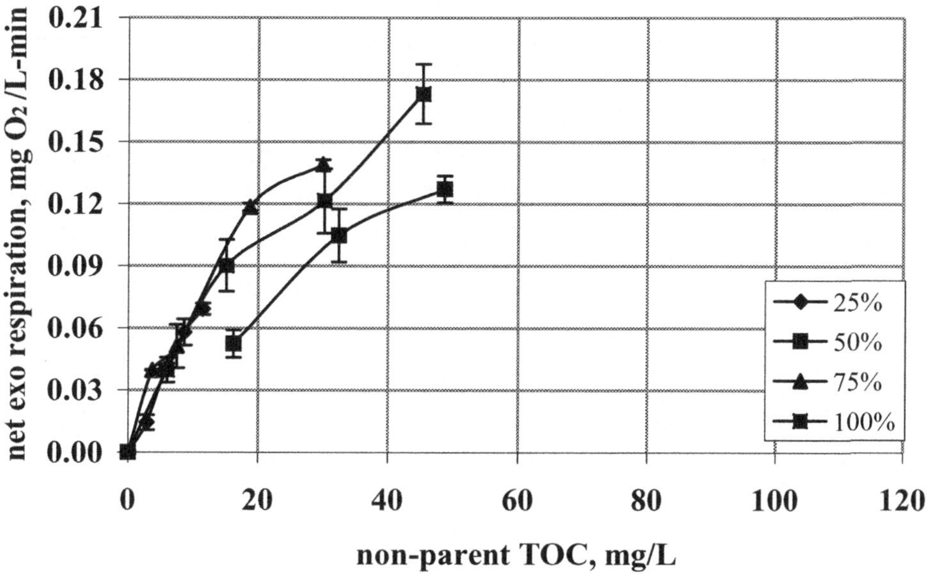

Figure 4-6. Respiration Rates of Non-Acclimated Cultures for UV/H_2O_2-Treated DCDE.

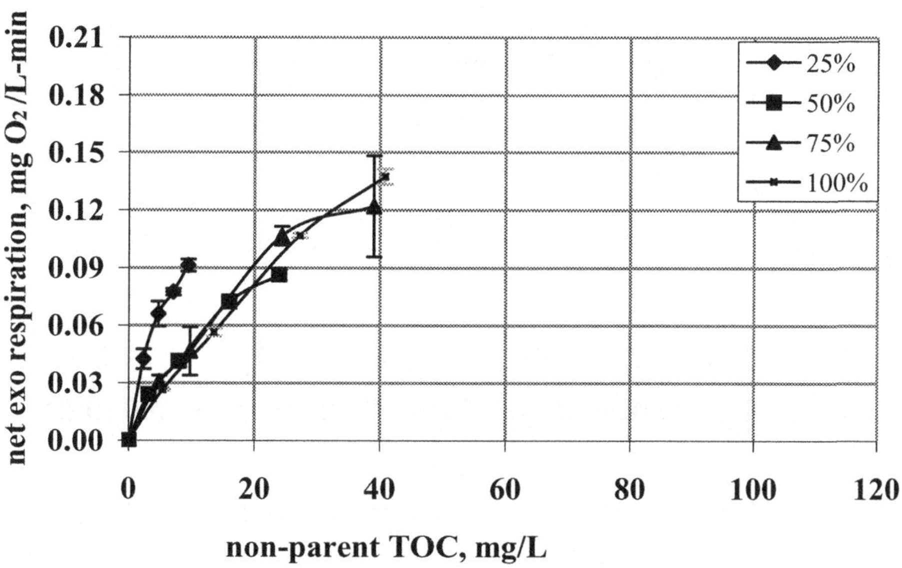

Figure 4-7. Respiration Rates of Acclimated Cultures for UV/H_2O_2-Treated DCDE.

The results presented in these figures are not significantly different from each other in terms of the responses of acclimated versus non-acclimated cultures, as well as the responses of the cultures to varying degrees of oxidation. The net respiration rate increased with increasing concentration of non-parent TOC, as expected, and as observed for the other oxidation methods.

4.1.4 Comparison of Treatments

The Monod maximum utilization rate estimated for ozone-, Fenton-, and UV/H_2O_2-treated DCDE solutions is presented in Figure 4-8 for comparison. Overall, the max respiration rates somewhat increased with increasing percent oxidation, indicating that slightly more biodegradable types of organic compounds are produced with increasing oxidation levels. Furthermore, the expected μ_{max} for the pre-oxidized DCDE ranged approximately between 0.1–0.2 mg oxygen L^{-1} min^{-1}, regardless of the oxidation method and acclimation or non-acclimation of the culture.

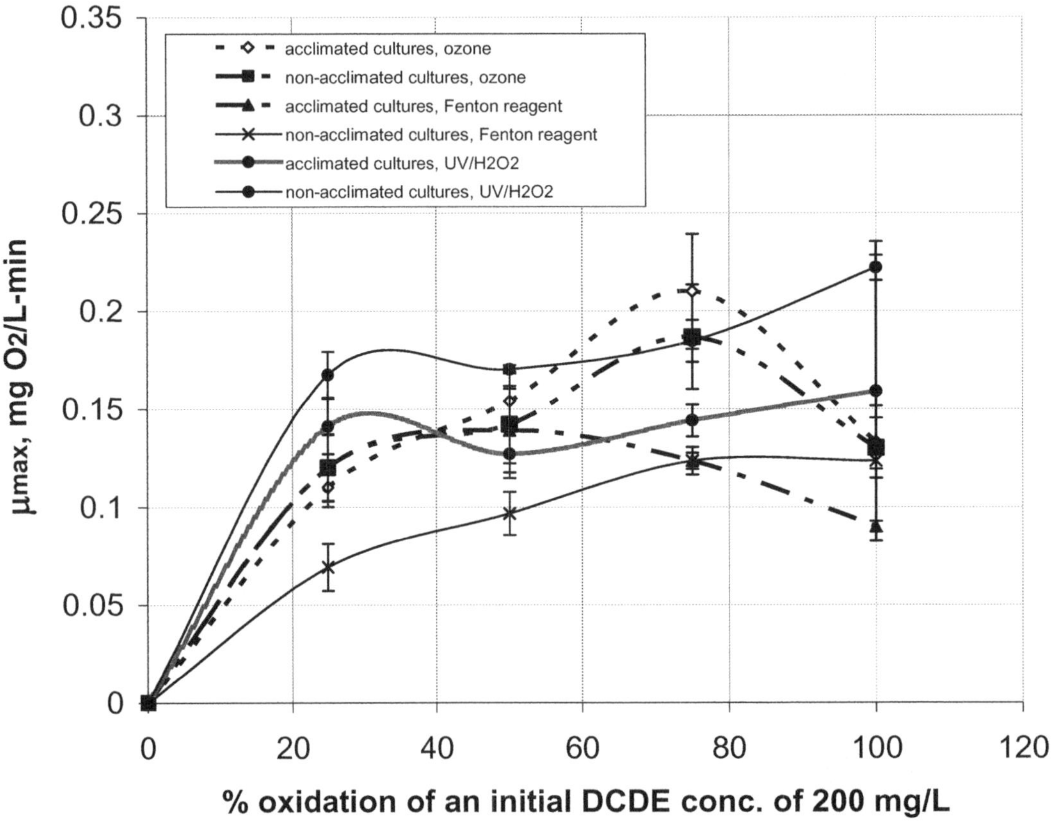

Figure 4-8. Comparison of max for Different Oxidation Methods and Acclimated and Non-Acclimated Cultures.

Comparison of the Monod half-saturation constant, K_s, for all treatments is presented in Figure 4-9. The three treatment methods produced significantly different K_s values. However, acclimation of the culture did not seem to have any significant effect for any of the treatment techniques. Although the data imply that Fenton-treated samples have more readily biodegradable by-products of oxidation, i.e., the lowest K_S, the presence of toxic by-products seemed to have resulted in a lower maximum respiration rate when compared to ozonation and UV/H_2O_2. Overall, the results presented in Figure 4-9 imply steeper slopes for oxygen consumption, lower K_s, for Fenton-treated samples than those treated by UV/H_2O_2. Ozone-treated samples had slopes in between the two treatments. The results are significantly different from each other, especially at the maximum oxidation level of 100%.

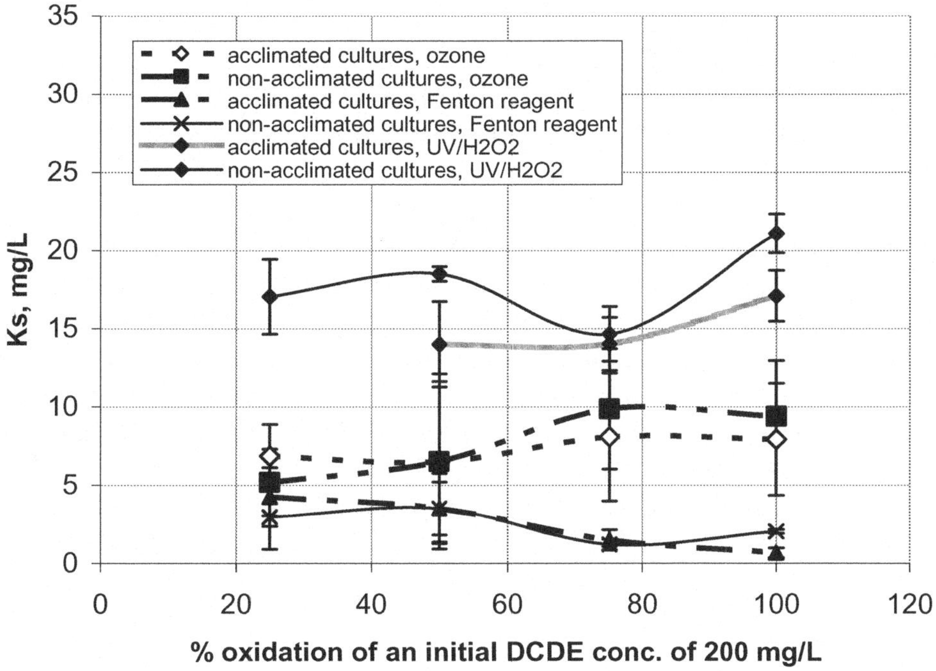

Figure 4-9. Comparison of K_s for Different Oxidation Methods and Acclimated and Non-Acclimated Cultures.

4.2 Long-Term Biodegradation Tests

4.2.1 Ozone-Treated DCDE

The results of the long-term biodegradation experiments for ozonated DCDE solutions are shown in Figure 4-10 for the non-acclimated culture. The figures show the overall oxygen decline and overall carbon dioxide formation in test solutions as a function of time. For the non-acclimated culture, the decline in oxygen concentration, i.e., the oxygen consumption, was slightly higher for the endogenous control than the oxygen consumption of non-oxidized DCDE. This observation indicates that DCDE was toxic at 100 mg/L. However, for the preoxidized DCDE samples, the oxygen decline was significantly higher than the endogenous control, resulting in an overall oxygen consumption of about 15 mg within 14 days for the two highly oxidized samples, which is equivalent to 45 mg/L of treated samples. Figure 4-11 shows the respirometric results of ozonated DCDE solutions with the acclimated culture. Here, the acclimated culture seemed to tolerate the presence of 100 mg/L of DCDE well initially but exhibited reduced oxygen consumption after about six days. The ozonated DCDE solutions consumed more oxygen compared to the endogenous respiration, and the oxygen consumption increased as the oxidation level increased.

4.2.2 Fenton-Treated DCDE

Fenton treatment also caused considerable oxygen decline and carbon dioxide production in the oxidized solutions compared to the non-oxidized DCDE (Figures 4-12 and 4-13). However, the improvement in oxygen decline and carbon dioxide production in the Fenton-treated samples ceased to exceed those values exhibited by endogenous respiration alone after about 40 hours (1.5 days) for non-acclimated and 150 hours (6 days) for the acclimated culture.

4.2.3 UV/H_2O_2-Treated DCDE

The results for UV/H_2O_2-treated DCDE are presented in Figures 4-14 and 4-15. Again, pre-oxidized solutions showed significantly higher metabolic activity compared to endogenous respiration alone.

4.2.4 Comparison of Treatments

The net oxygen decline and net carbon dioxide production for the three oxidation methods for both acclimated and non-acclimated cultures are presented in Figure 4-16 for the 2-week experiments. The net oxygen decline was calculated by subtracting the oxygen consumption of endogenous control, i.e., metabolic activity of microorganism alone, from the consumption by microorganisms in the presence of the oxidized solutions. Similarly, carbon dioxide production of the endogenous control was subtracted from the carbon dioxide production observed in samples containing oxidized solutions to obtain the net values. The results are presented for solutions that were oxidized at highest oxidation levels (90–100% oxidation) for all three oxidation processes.

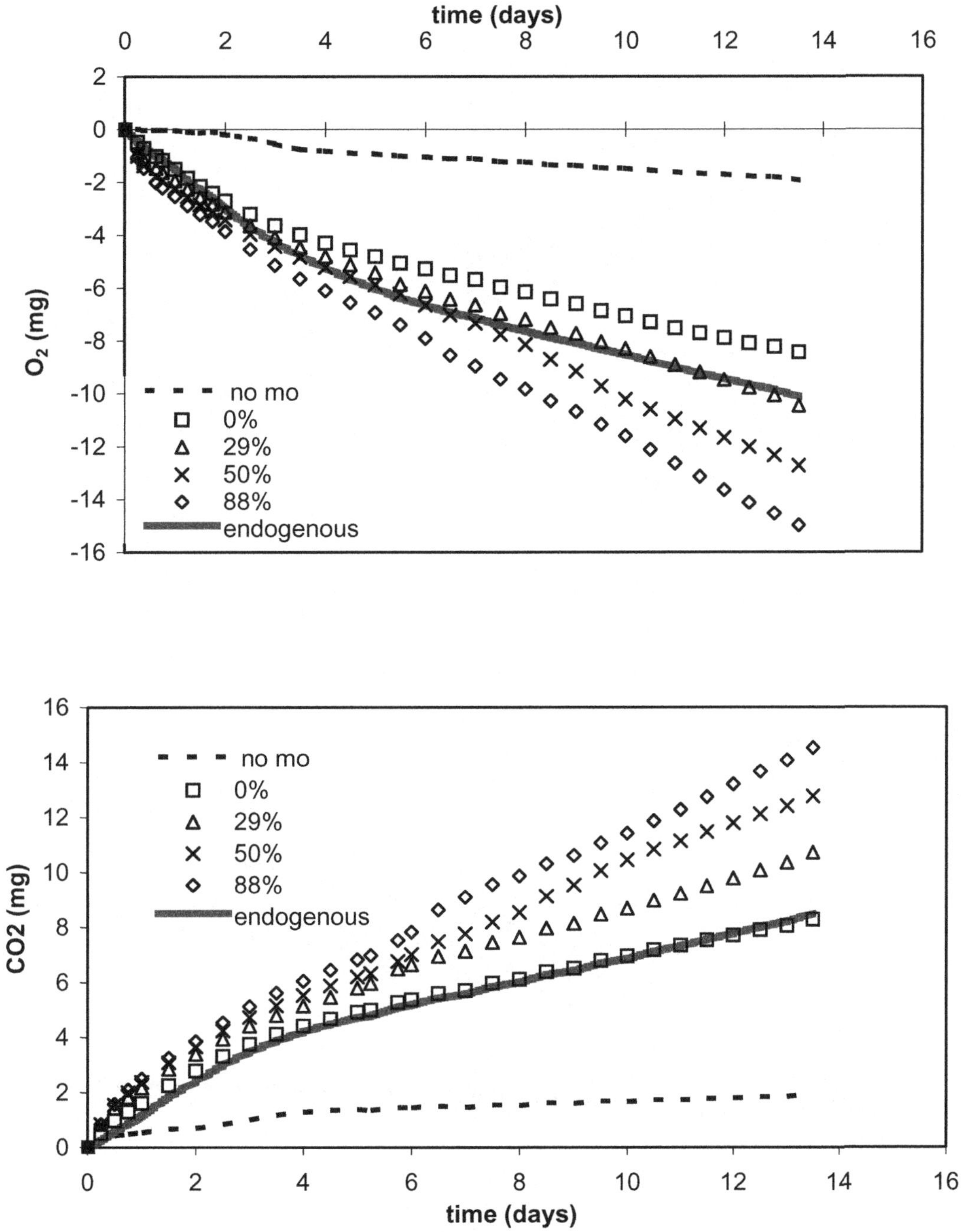

Figure 4-10. Oxygen Decline and Carbon Dioxide Production with Non-Acclimated Culture of Ozone-Treated DCDE.

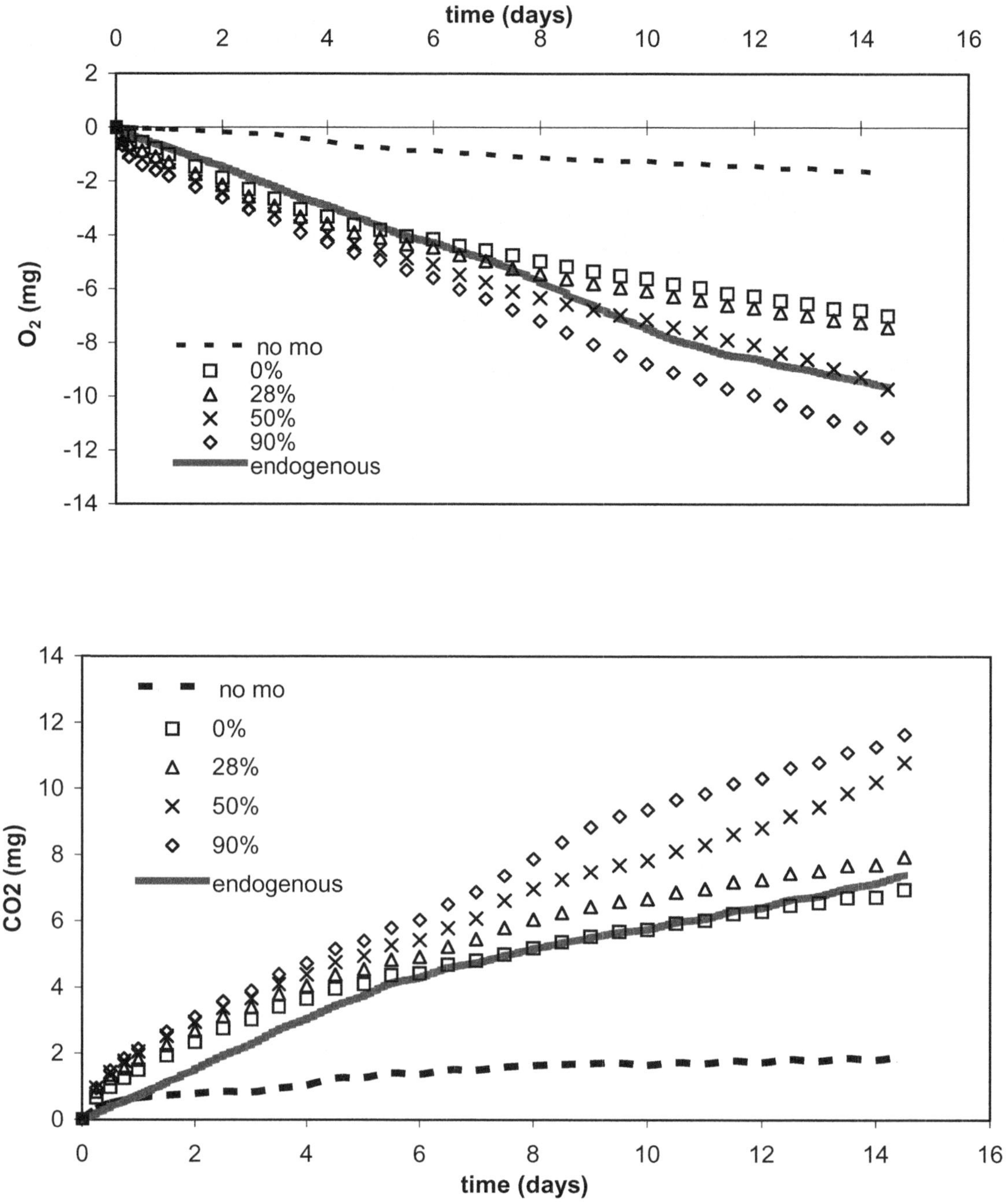

Figure 4-11. Oxygen Decline and Carbon Dioxide Production with Acclimated Culture of Ozone-Treated DCDE.

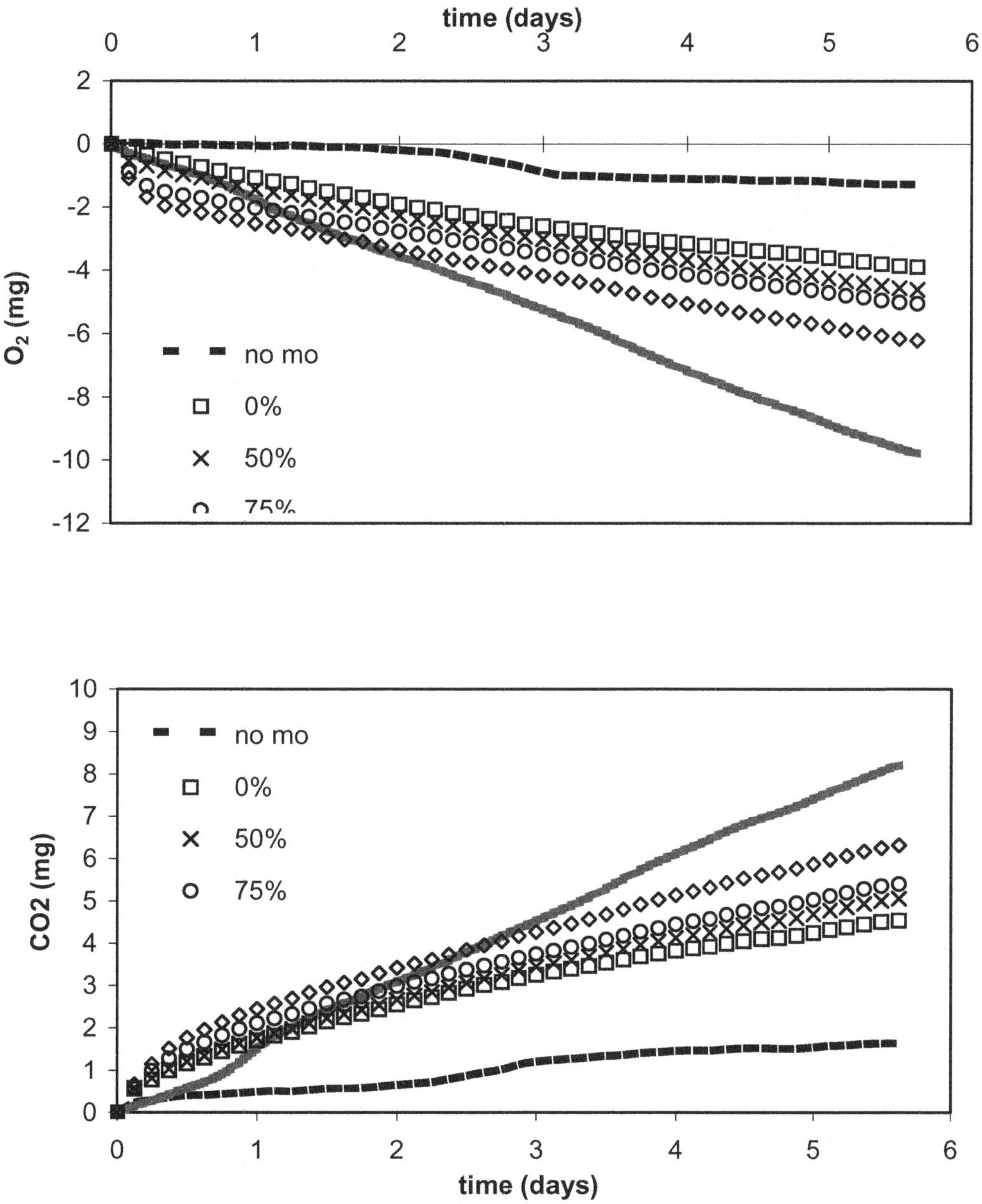

Figure 4-12. Oxygen Decline and Carbon Dioxide Production with Non-Acclimated Cultures of Fenton-Treated DCDE.

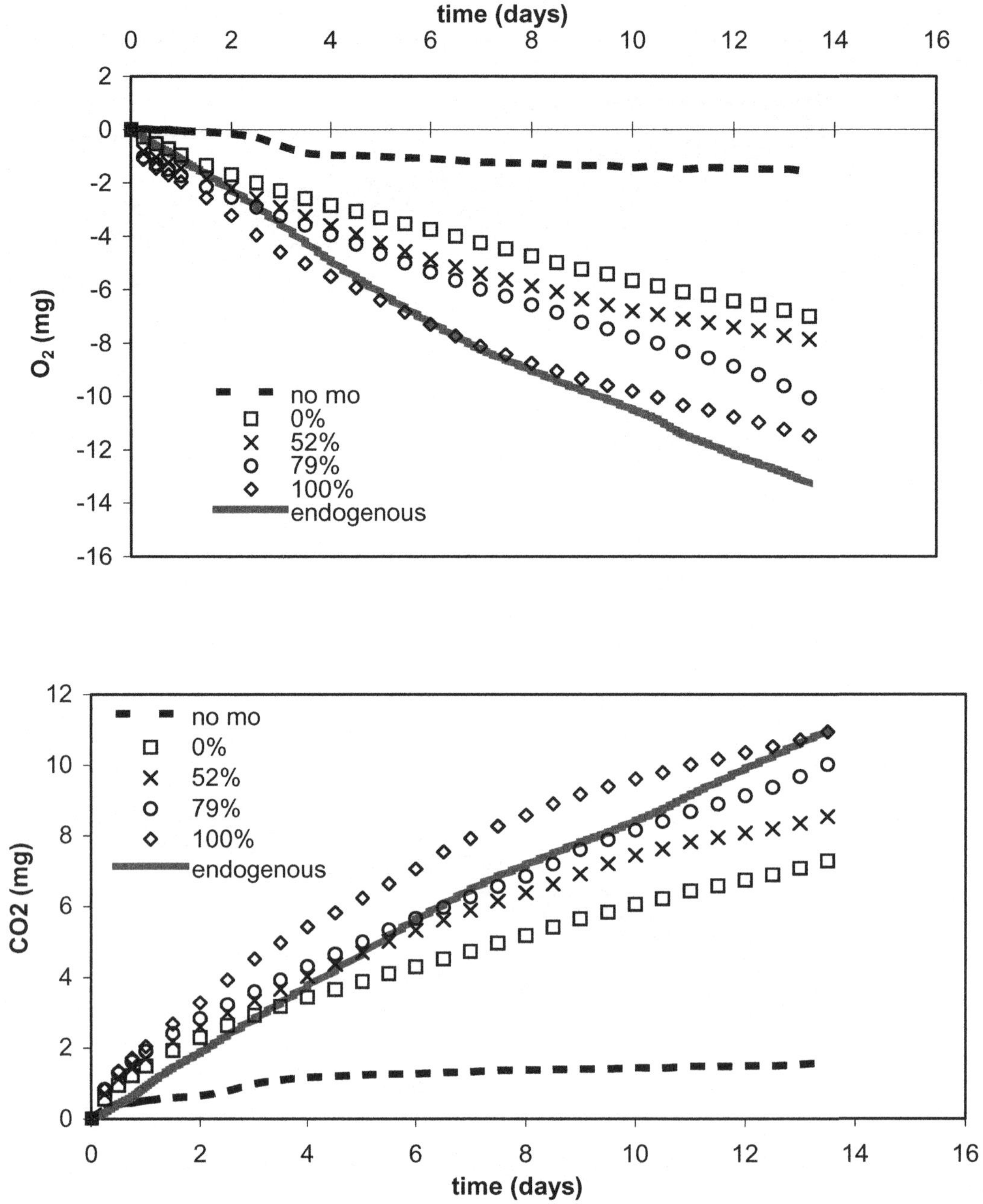

Figure 4-13. Oxygen Decline and Carbon Dioxide Production with Acclimated Cultures of Fenton-Treated DCDE.

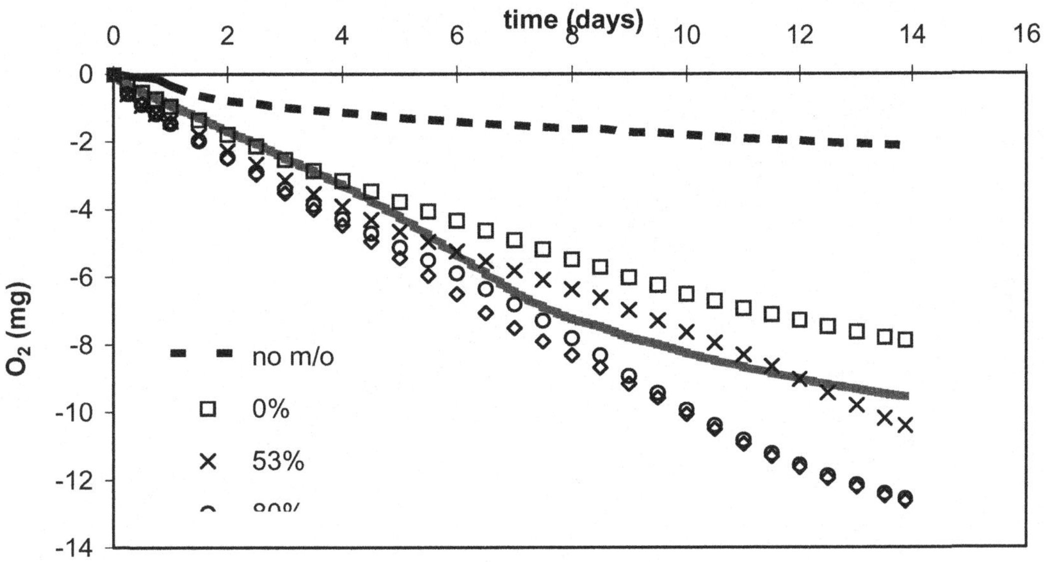

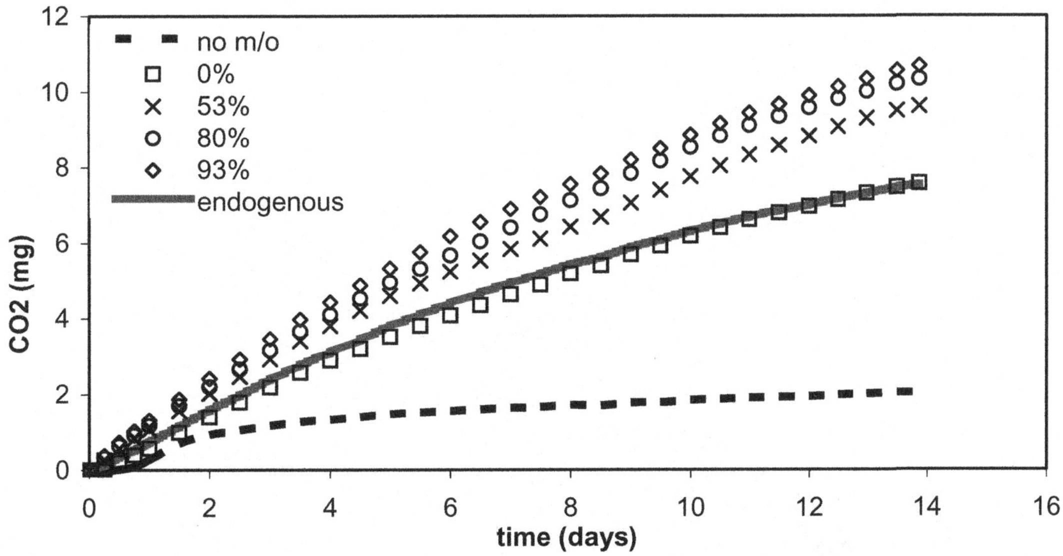

Figure 4-14. Oxygen Decline and Carbon Dioxide Production with Non-Acclimated Culture of UV/H_2O_2-Treated DCDE.

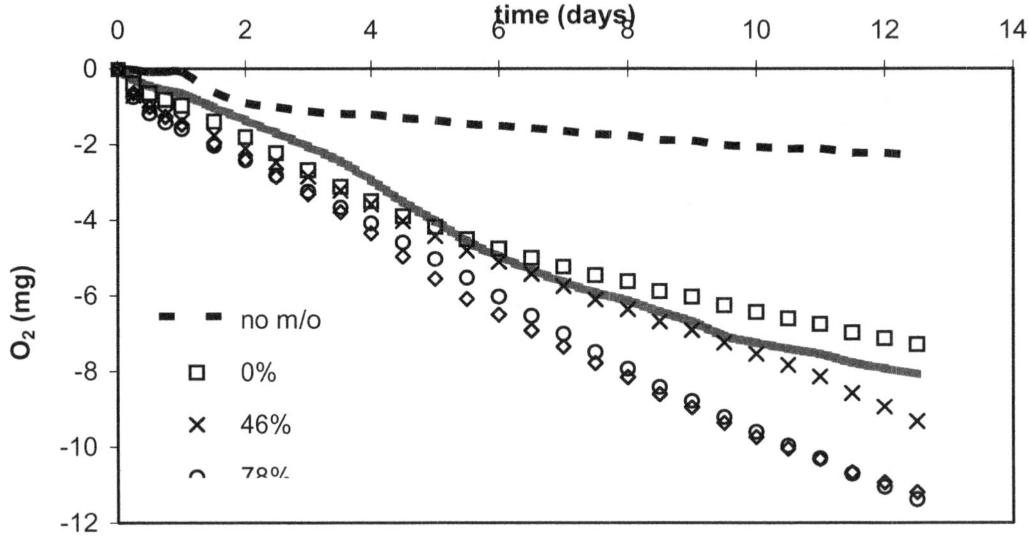

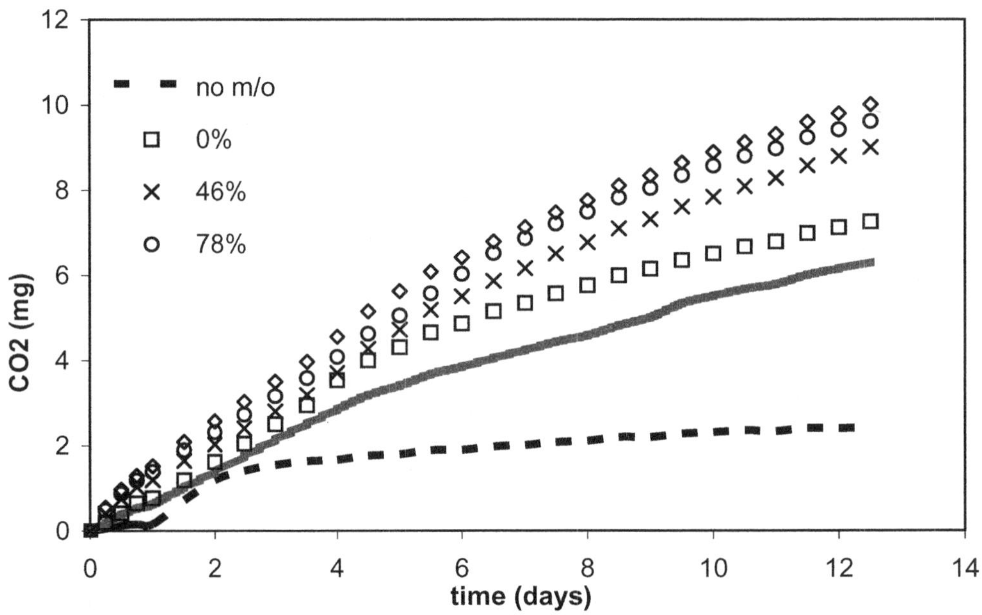

Figure 4-15. Oxygen Decline and Carbon Dioxide Production with Acclimated Culture of UV/H_2O_2-Treated DCDE.

WERF

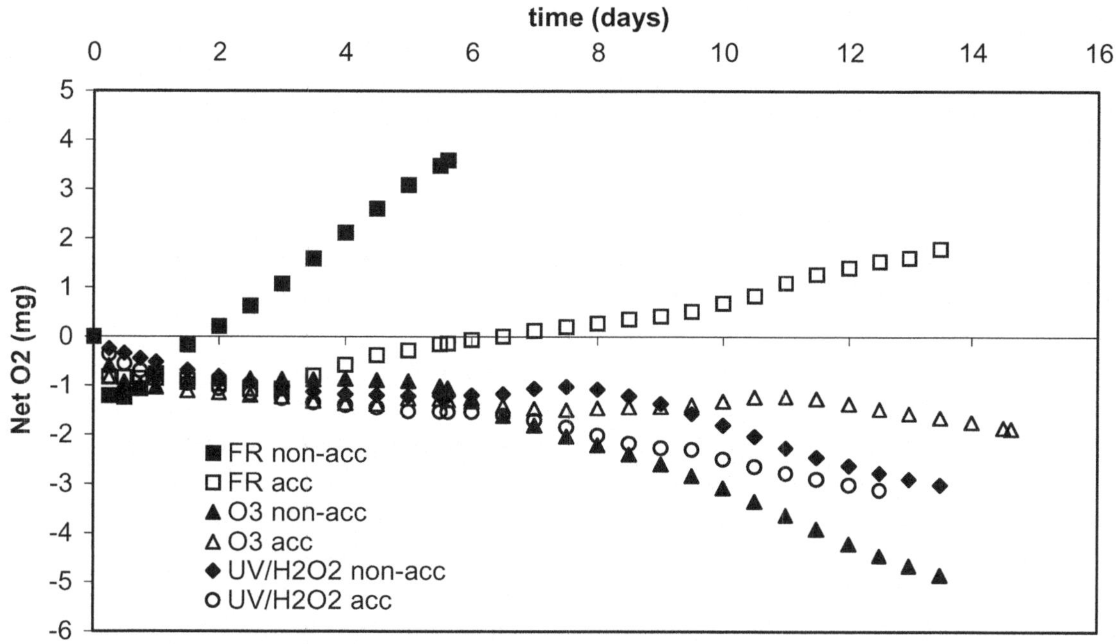

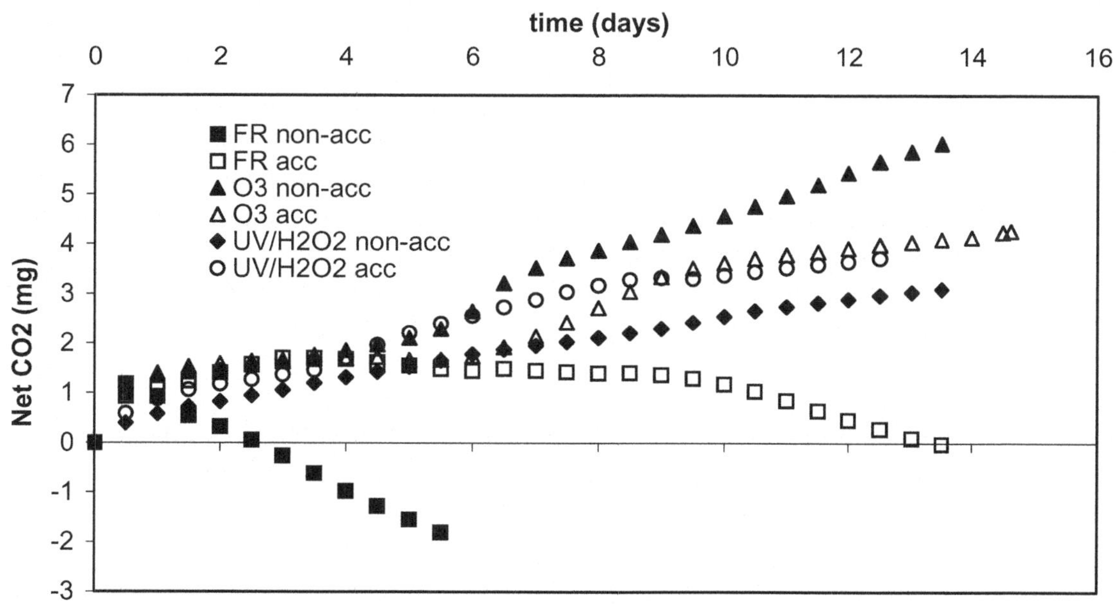

Figure 4-16. Comparison in Net Oxygen Decline and Carbon Dioxide Production of Fenton, Ozone, and UV/H_2O_2-Treated Samples Using Acclimated and Non-Acclimated Cultures. "Net" indicates sample value minus endogenous value.

The Fenton-treated samples are depicted by the squares in these figures. After the initial period of about 70 hours for acclimated cultures (open squares) and only 10 hours for non-acclimated cultures (full squares), oxygen decline was lower than the endogenous control, reversing the slope, and eventually resulting in higher than zero values for net oxygen levels. The trend was the same for carbon dioxide production. These results indicate that Fenton reagent produced highly toxic by-products that have inhibited the metabolic activity of the microorganisms. The non-acclimated culture showed much more sensitivity to these products than the acclimated culture.

DCDE samples oxidized by ozone and UV/H_2O_2 continued to have increasing metabolic activity, although ozone-treated samples exhibited a significantly higher metabolic rate with non-acclimated cultures compared to the others.

It should be noted that the short-term tests, which were conducted over a period of 30 minutes, did not reveal any significant differences in μ_{max} values among the three oxidation methods. No toxicity was apparent in Fenton-treated samples. In fact, steeper slopes for metabolic activity for Fenton-treated samples were observed. This is understandable based on the data shown in Figure 4-17 for the first 24 hours of the 2-week tests. All the samples, including those treated by Fenton reagent, showed considerable oxygen decline, i.e., oxygen consumption, and carbon dioxide production during the first 5 hours of the test. That is consistent with the results of the short-term tests. In fact, Fenton-treated samples seemed to exert higher metabolic rates initially, creating steeper slopes for oxygen decline and carbon dioxide formation during the first five hours. This is consistent with the results of the short-term tests.

Additional observations on results presented in Figures 4-10 through 4-15 include the following:

Higher oxidation levels led to higher metabolic rates and extent during the respirometric testing for all three oxidation methods. Furthermore, oxygen data were found more sensitive than CO_2 data to the treatments. For example, all ozone and UV/H_2O_2 data indicated that based on CO_2 formation, all levels of oxidation caused respiration higher than the endogenous respiration. However, oxygen data show that samples oxidized by ozone at 29% level started to show lower respiration after two days for non-acclimated culture. Similarly, for the UV/H_2O_2 data the non-acclimated culture exhibited lower respiration for the 53% and 23% oxidized samples than the endogenous respiration after four and six days.

In addition, the data indicate that for both ozonated and UV/H2O2-treated samples, non-acclimated culture showed more tolerance than the acclimated cultures to the partially oxidized DCDE samples, i.e., those that contained considerable DCDE.

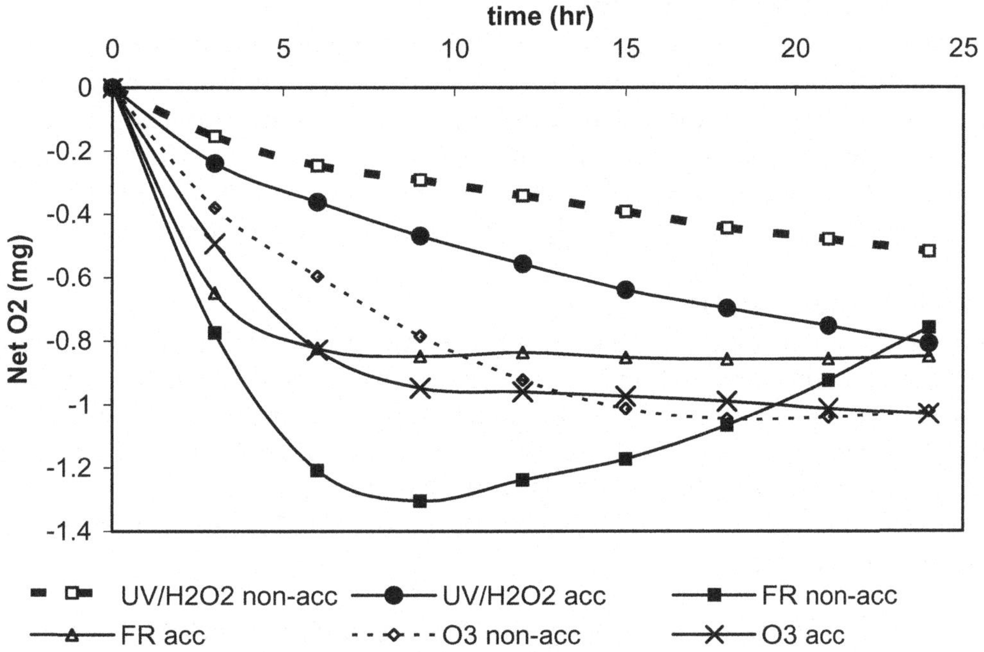

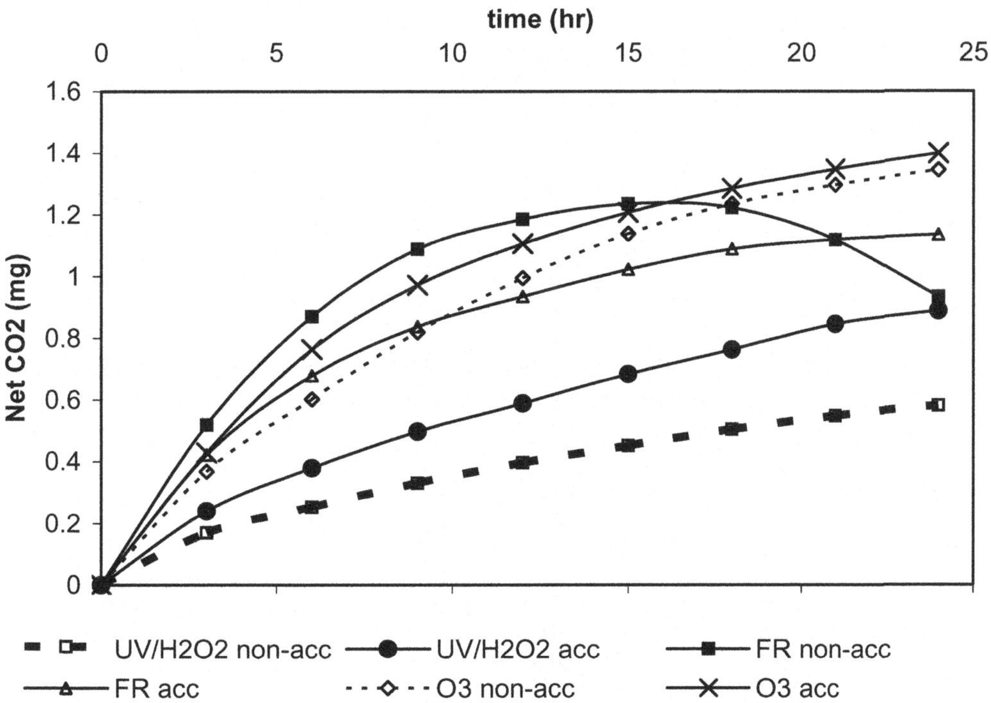

Figure 4-17. Comparison of Net Oxygen Decline and Carbon Dioxide Production of Fenton-, Ozone-, and UV/H_2O_2-treated Samples Using Acclimated and Non-Acclimated Cultures during the First 24 hours. "Net" indicates sample value minus endogenous value.

4.3 Mid-Term Biodegradation Tests

The mid-term biodegradation experiments were conducted with DCDE solutions preoxidized at 75% and 100%. These two levels of oxidation were chosen because of the higher rate of respiration obtained for these levels with the short and long term tests. In fact, industry might need to oxidize recalcitrant chemicals by 100%, otherwise, the effluent stream could still contain concentrations that would exceed permit limits, especially since these recalcitrant compounds are expected to pass through the treatment sequence without being removed.

4.3.1 Ozone-Treated DCDE

The experimental results for mid-term biodegradation experiments for an ozonated solution are presented in Table 4-1 and Table 4-2 for non-acclimated and acclimated cultures, respectively. The results presented in the following Tables are averages of duplicate experiments and presented as changes in concentrations of non-parent COD, non-parent TOC, and DCDE concentration during a 24-hour period. DCDE solution at an initial concentration of 200 mg/L was pre-ozonated to reduce DCDE by 94%. All the preozonated solutions were diluted by 50% for the biodegradation tests.

Table 4-1. Biodegradation of 94% Oxidized DCDE by Ozone. Non-Acclimated Culture. (The results are in non-parent COD and TOC)

Time (hr)	COD (mg/L)	TOC (mg/L)	DCDE (mg/L)
0.0	79.4	38.1	5.9
1.5	68.9	32.9	5.5
3.0	65.0	31.0	5.2
4.5	66.7	30.7	4.8
6.0	60.5	29.2	4.5
21.0	58.5	27.5	0
24.0	61.0	27.9	0
Δ24	18.4	11.1	5.9

Table 4-2. Biodegradation of 94% Oxidized DCDE by Ozone. Acclimated Culture. (The results are in non-parent COD and TOC)

Time (hr)	COD (mg/L)	TOC (mg/L)	DCDE (mg/L)
0.0	81.7	37.0	5.9
1.5	65.9	32.4	5.0
3.0	64.1	30.8	4.4
4.5	58.8	30.0	3.9
6.0	60.3	27.6	3.8
21.0	56.9	26.3	0
24.0	51.5	28.2	0
Δ24	30.2	8.8	5.9

Based on Δ24 COD values, it can be concluded that the acclimated culture performed better than non-acclimated culture for 94%-ozonated samples. On the other hand, based on ΔTOC values, there was no significant difference between the two cultures. DCDE concentration reduced slightly over the first 6 hours of the experiment. However, at the end of 24 hours, no DCDE was detected in the solutions.

Similar experiments were also conducted on DCDE samples ozonated at the 73% removal levels. The results will be presented in summary form in Section 4.3.4.

4.3.2 Fenton-Treated DCDE

The experimental results for mid-term biodegradation experiments for a Fenton-treated DCDE solution at 90% removal level are presented in Table 4-3 and Table 4-4 for non-acclimated and acclimated cultures, respectively.

Table 4-3. Biodegradation of 90% Oxidized DCDE by Fenton Reagent Non-Acclimated Culture. (The results are in non-parent COD and TOC)

Time (hr)	COD (mg/L)	TOC (mg/L)	DCDE (mg/L)
0.0	76.4	26.7	11.6
1.5	73.9	23.8	10.3
3.0	65.3	22.8	9.6
4.5	63.1	21.7	9.1
6.0	55.1	20.5	8.7
12.0	54.0	18.8	7.7
18.0	47.3	18.2	6.5
24.0	47.3	18.5	5.6
Δ24	29.1	8.2	6.0

Table 4-4. Biodegradation of 90% Oxidized DCDE by Fenton Reagent Acclimated Culture. (The results are in non-parent COD and TOC)

Time (hr)	COD (mg/L)	TOC (mg/L)	DCDE (mg/L)
0.0	82.0	28.4	11.4
1.5	62.6	22.3	10.2
3.0	50.4	19.4	9.3
4.5	52.4	19.3	8.3
6.0	47.5	17.5	7.5
12.0	53.0	19.1	5.5
18.0	50.6	18.1	3.7
24.0	54.6	19.7	2.5
Δ24	27.4	8.7	8.9

Similar experiments were also conducted on DCDE samples Fenton-treated at the 72% removal levels. The results will be presented in summary form in Section 4.3.4.

4.3.3 UV/H_2O_2-Treated DCDE

The experimental results for mid-term biodegradation experiments for a UV/H_2O_2 DCDE solution at 100% removal level are presented in Table 4-5 and Table 4-6 for non-acclimated and acclimated cultures, respectively.

Table 4-5. Biodegradation of 100% Oxidized DCDE by UV/H_2O_2 Non-Acclimated Culture.

Time (hr)	COD (mg/L)	TOC (mg/L)
0.0	77.0	35.7
1.5	77.0	34.0
3.0	75.0	32.0
4.5	71.0	27.5
6.0	62.0	26.5
21.0	50.0	27.3
24.0	48.5	-
	$\Delta_{24} = 28.5$	$\Delta_{21} = 8.4$

Table 4-6. Biodegradation of 100% Oxidized DCDE by UV/H_2O_2 Acclimated Culture.

Time (hr)	COD (mg/L)	TOC (mg/L)
0.0	87.5	32.5
1.5	75.0	31.9
3.0	73.0	32.5
4.5	64.0	29.3
6.0	69.0	23.7
21.0	66.0	22.0
24.0	60.0	-
Δ_{24}	$\Delta_{24} = 27.5$	$\Delta_{21} = 10.5$

4.3.4 Comparison of Treatments

The removal of npCOD and npTOC based on the average and range of data of the duplicate tests are compared in Table 4-7 for different treatments over the *first 6 hours* of treatment. Note that the measured values for percent oxidation of DCDE varied from the target values of 75% and 100% considerably, and the actual values are reported in the table.

Table 4-7. Comparison of Treatments on the Basis of Overall Removal of Organic Matter (npCOD and npTOC) in Terms of Average Values and Range of Data of Duplicate Tests During *6 hours* of Biodegradation Experiments.

	$npCOD_{AVG,}$ *mg/L,* REMOVED	$npTOC_{AVG}$*, mg/L,* REMOVED
High oxidation levels		
Ozone, 94%, acclimated culture	25.1 (19.7 – 30.5)	9.5 (4.4-14.6)
Ozone, 94%, non-acclimated culture	19.0 (14.2 – 23.8)	8.2 (6.5 – 9.8)
Fenton reagent, 90% acclimated culture	35.0 (28.8 – 41.2)	10.9 (10.8 – 11)
Fenton reagent, 90% non-acclimated culture	21.6 (18.8 – 24.5)	6.2 (4.6 – 7.8)
UV/H_2O_2, 100% acclimated	18.5 (8.0 - 29.0)	8.8 (4.4- 13.2)
UV/H_2O_2, 100% non-acclimated	15.0 (12.0 – 18.0)	9.2 (4.7 – 13.7)
Low oxidation levels		
Ozone, 73%, acclimated culture	22.7 (20.4 –25.0)	4.9 (4.4 – 5.3)
Ozone, 73%, non-acclimated culture	26.5 (18.3 – 34.6)	3.9 (3.1 – 4.7)
Fenton reagent, 72%, acclimated culture	26.4 (23.0 – 29.8)	5.1 (4.7 – 5.4)
Fenton reagent, 72%, non-acclimated culture	22.6 (18.0 – 27.2)	4.7 (4.4 – 5)

About 23 ± 12 mg/L of non-parent COD was removed over the 6-hour period for ozonated, Fenton-treated, and UV/H_2O_2–treated DCDE at both high and low oxidation levels. This is about 20% of the theoretical initial COD value (the theoretical COD value of 100 mg/L of the original DCDE is 112 mg/L). The treatments do not seem to be very different from one another in terms of npCOD removal efficiencies. Likewise, non-acclimated cultures and acclimated cultures performed about the same under these conditions.

In terms of mineralization of organic matter, about 9 mg/L (8.8 ± 2.6) of npTOC biodegraded during the 6-hour test period for the "high oxidation" samples, i.e., those oxidized by 90% or more. The npTOC removals among the first four samples are not significantly different from each other after taking into consideration the errors involved in the TOC measurements. However, there is a noticeable difference in the removal of the npTOC between the high and low oxidation levels for the treatments. On average, after low oxidation levels (72%–73%), the microorganisms were able to remove only about 5 mg/L (4.6± 1.2) of npTOC.

These results indicate that higher oxidation levels produce higher concentrations of the by-products that are more readily biodegradable, and thus lead to the mineralization of a larger fraction of the initial organic matter. This agrees with expectations that higher levels of chemical oxidation would lead to formation of more oxygenated and smaller weight products, e.g., carboxylic acids, that can easily be converted to CO_2 by microorganisms.

However, even for high oxidation levels, the percentage of TOC removal is only 25% of the initial theoretical TOC (the theoretical TOC value of 100 mg/L of DCDE is 34 mg/L). In fact, chemical oxidation was not able to remove any TOC since the average of the TOC values listed in Tables 5-1 through Tables 5-6 at the initial time is 33 mg/L. The average TOC removal of 9 mg/L during 6-hours of the biodegradation experiment indicate that only 25% of the original organic matter present in the solutions after chemical oxidation biodegraded under the described conditions. This relatively low organic matter removal may be attributed either to the formation of limited amounts of biodegradable oxidation products and/or to formation of oxidation products that are inhibitory and/ or toxic for the types of microorganisms present in the culture. The results of the long-term biodegradation tests showed the toxicity for the Fenton-treated samples. However, no toxicity was observed for the ozonated samples. On the other hand, it can be hypothesized that higher organic matter removal may be obtained in activated sludge systems that contain at least five to six times higher concentrations of microorganisms. Furthermore, longer exposures of the biomass might facilitate acclimation of the biomass to the oxidation products.

CHAPTER 5.0

Assessment of Biodegradation of MTBE

5.1 Short-Term Biodegradation Test

The experimental protocol and procedure for data analysis presented earlier for short-term biodegradation studies for DCDE were followed here for MTBE. The results for acclimated and non-acclimated cultures for all three oxidation methods are presented below.

5.1.1 Ozone-Treated MTBE

The net exogenous respiration rates for acclimated and non-acclimated cultures are presented in Figure 5-1. Here, the abscissa is presented in non-parent TOC concentration.

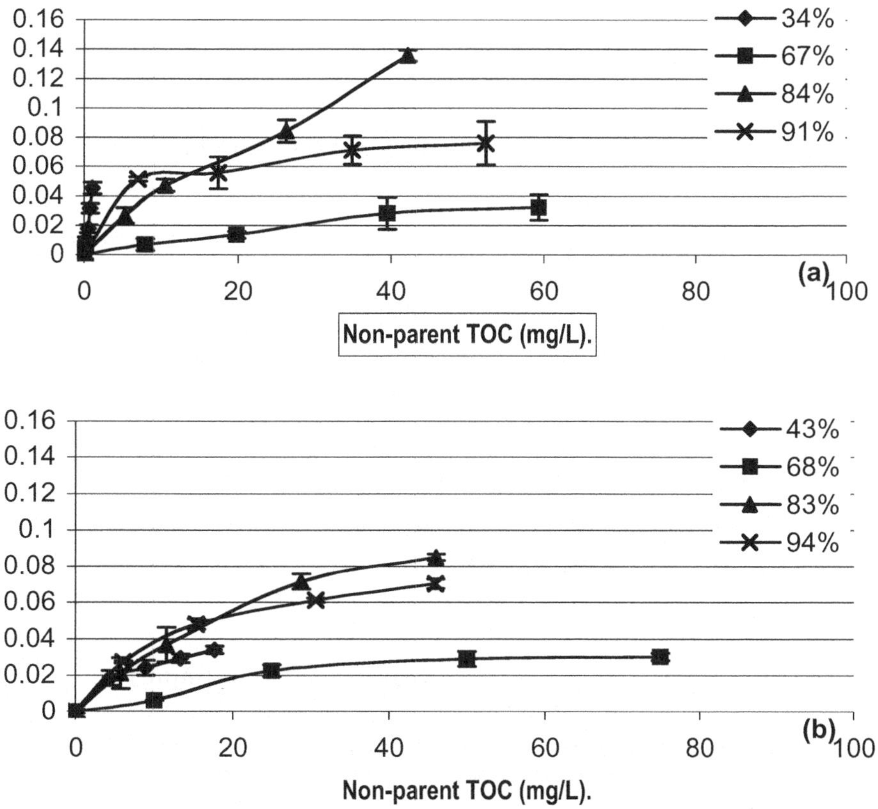

Figure 5-1. Biodegradability of Ozone-Treated MTBE with (a) Acclimated and (b) Non-Acclimated Cultures with Abscissa in Non-Parent TOC Concentration.

The bacterial respiration rate increased with increasing oxidation levels for both acclimated and non-acclimated cultures, showing that ozonation produced more readily biodegradable oxidation products. The respiration rate was found higher at 83%–84% oxidation level than at 91–94% oxidation level for both cultures. A similar behavior was also observed for cultures biodegrading oxidized DCDE. Significant differences between the TOC of MTBE solution before and after ozonation were noted, and the TOC change was attributed primarily to the stripping of MTBE during ozonation.

5.1.2 Fenton-Treated MTBE

The net exogenous respiration rates for acclimated and non-acclimated cultures are presented in Figure 5-2, with the abscissa presented in non-parent TOC concentration.

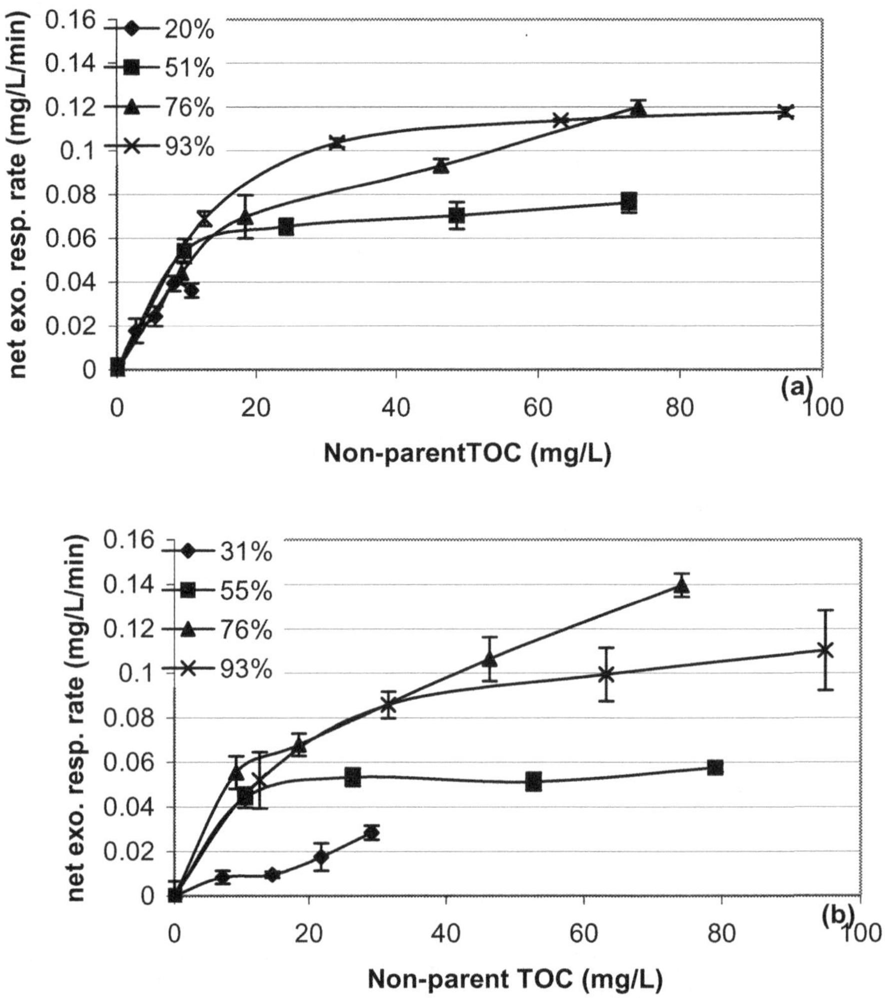

Figure 5-2 Biodegradability of Fenton-treated MTBE by (a) Acclimated and (b) Non-Acclimated Cultures with Abscissa in Non-Parent TOC Concentration.

The results obtained with Fenton oxidation are very similar to those obtained with ozone oxidation for both cultures. There is no significant difference between the maximum respiration rates obtained for the two cultures. The highest respiration rates for samples oxidized with Fenton reagent were observed at 76% and 93% oxidation levels.

5.1.3 UV/H_2O_2 –Treated MTBE

The net exogenous respiration rates for acclimated and non-acclimated cultures are presented in Figure 5-3, with the abscissa presented in non-parent TOC concentration.

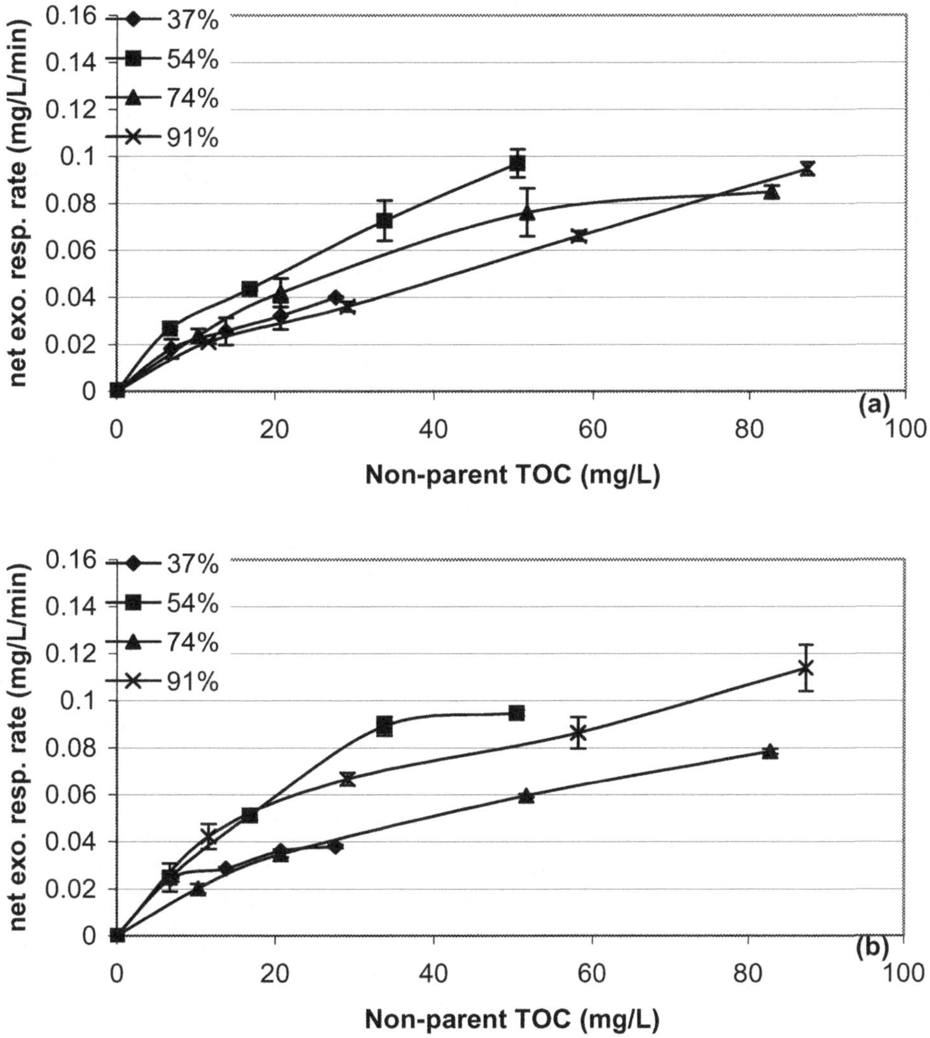

Figure 5-3 Biodegradability of UV/H_2O_2-treated MTBE by (a) Acclimated and (b) Non-Acclimated Cultures with Abscissa in Non-Parent TOC Concentration.

UV/H_2O_2–treated MTBE samples followed the same trend as ozone and Fenton-treated MTBE in terms of the increase in the respiration rates with increasing non-parent TOC levels. The respiration rates obtained with acclimated and non-acclimated cultures were not different from each other.

5.1.4 Comparison of Treatments

The maximum respiration rates obtained for samples treated by all three oxidation methods are presented in Figure 5-4.

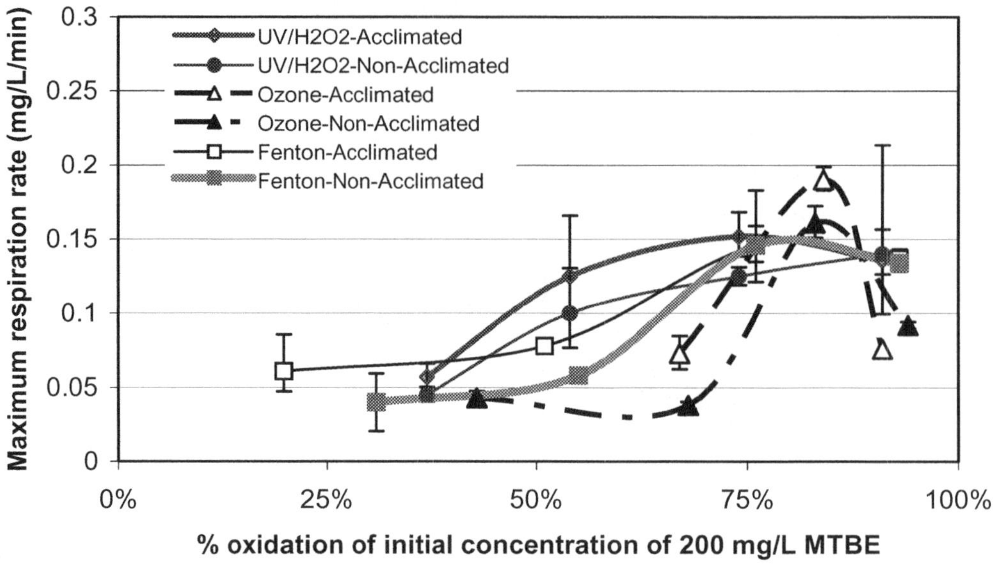

Figure 5-4. Comparison of μ_{max} After Ozone, Fenton, or UV/H_2O_2 Treatment of MTBE.

Overall, the biodegradability of MTBE solutions improved after chemical oxidation, indicating that the oxidation products were more biodegradable than MTBE.

Increasing oxidation levels produced higher respiration rates for all three oxidation methods. Furthermore, the maximum respiration rate observed for all three methods was about 0.15–0.2 mg/L-min. For ozone treatment, the μ_{max} values showed a peak at about 75%–80% oxidation level, and then reduced significantly for higher oxidation levels. A similar trend was observed for ozone-treated DCDE as well. Comparison of K_s values for different treatments was not presented, since it was difficult to extract any trend, and hence any meaningful information, from the data.

5.2 Long-Term Biodegradation Tests

The respirometric procedure described in earlier reports for DCDE was followed to study the biodegradability of MTBE as well. The oxidation levels tried for MTBE were zero (no oxidation) and approximately 50%, 75%, and 100% oxidation. The concentration of the biomass used in these studies was 400 mg/L for both acclimated and non-acclimated cultures, as before.

5.2.1 Ozone-Treated MTBE

The net decline in oxygen concentration and net production of carbon dioxide obtained by subtracting the values for endogenous respiration are plotted in Figures 5-5 and 5-6 for acclimated and non-acclimated cultures.

The results indicate that MTBE without preoxidation (0%) biodegraded readily with the acclimated culture but only slightly with the non-acclimated culture. It should be noted that during the short-term biodegradation studies, no biodegradation of MTBE was observed with acclimated or non-acclimated cultures.

In addition, pre-ozonation of MTBE at 60% and 80% levels did not produce any enhancement of biodegradation in non-acclimated culture. The net total oxygen consumption at the end of 14 days for MTBE oxidized at 0%, 60%, and 80% levels was about 0.5 mg. Whereas, the net oxygen consumption for MTBE ozonated at 97% level was about 3.8 mg. The trend was different for the acclimated culture. The net oxygen consumption for all treatments (0%–97% oxidation) was in the range of 4.5–5.5 mg at the end of 14 days, although the most highly oxidized sample (97%) resulted in the highest oxygen consumption and carbon dioxide production.

5.2.2 Fenton-Treated MTBE

The net decline in oxygen concentration and net production of carbon dioxide obtained by subtracting the values for endogenous respiration are plotted in Figures 5-7 and 5-8 for acclimated and non-acclimated cultures.

Again, MTBE without any oxidation showed considerable biodegradation with the acclimated culture and almost no degradation with the non-acclimated culture. Increasing levels of oxidation by Fenton reagent resulted in increased oxygen consumption and carbon dioxide production. The highest respiration rate was obtained for the sample oxidized at 94% level for both cultures. Overall, the acclimated culture produced higher respiration rates for partially oxidized MTBE.

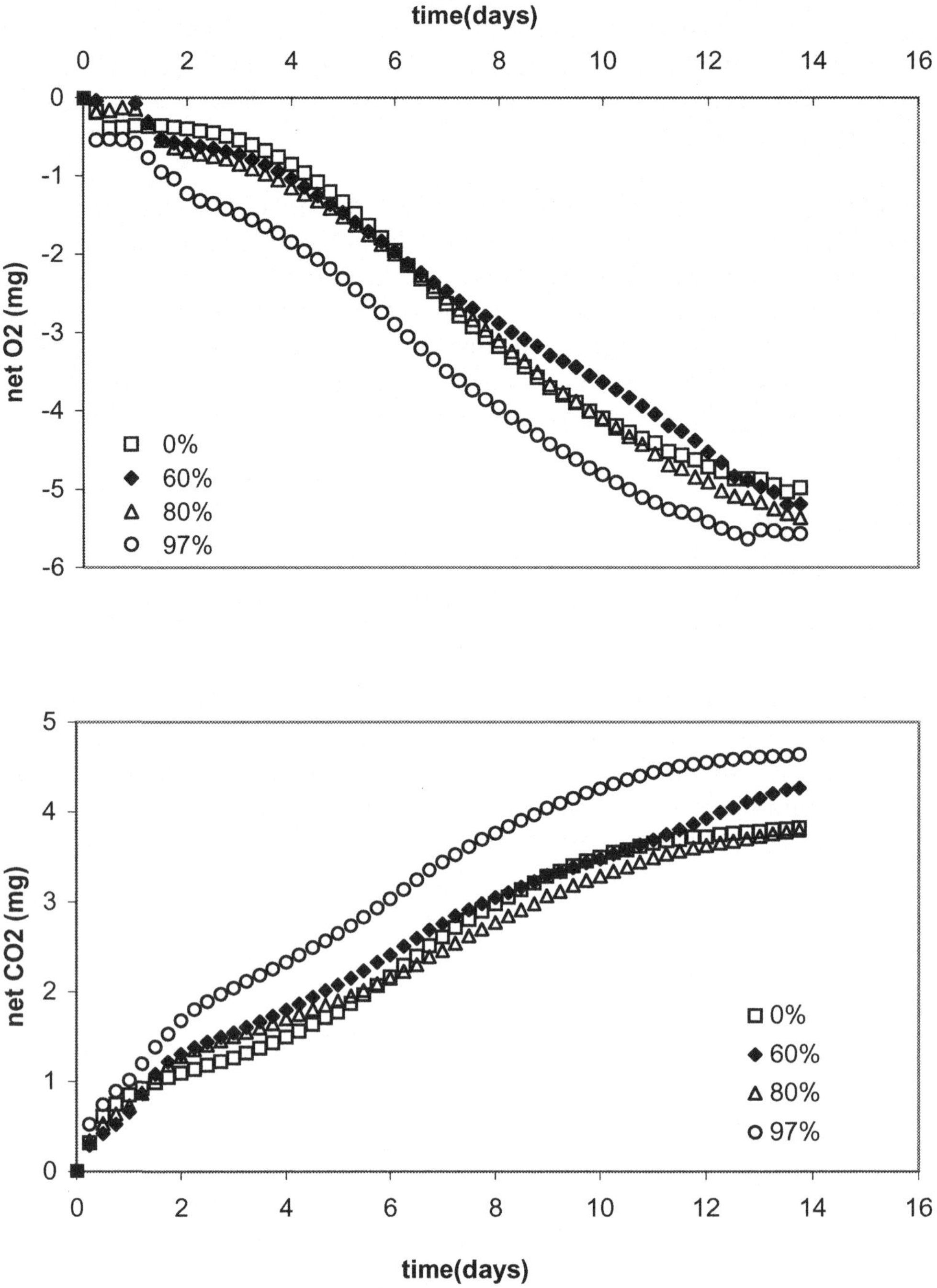

Figure 5-5. Net Oxygen Decline and Carbon Dioxide Production with Acclimated Culture of Ozone-Treated MTBE.

WERF

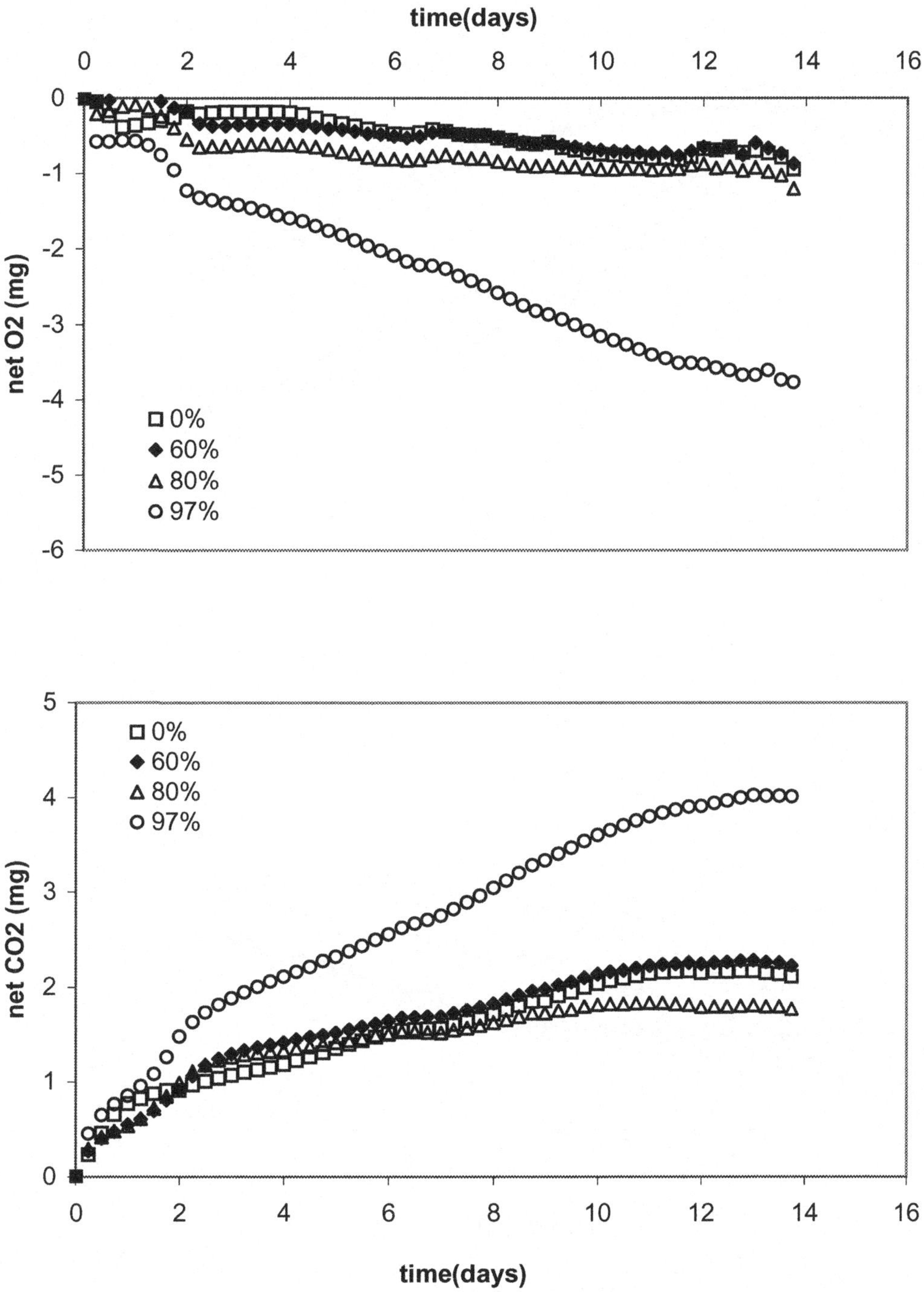

Figure 5-6. Net Oxygen Decline and Carbon Dioxide Production with Non-Acclimated Culture of Ozone-Treated MTBE.

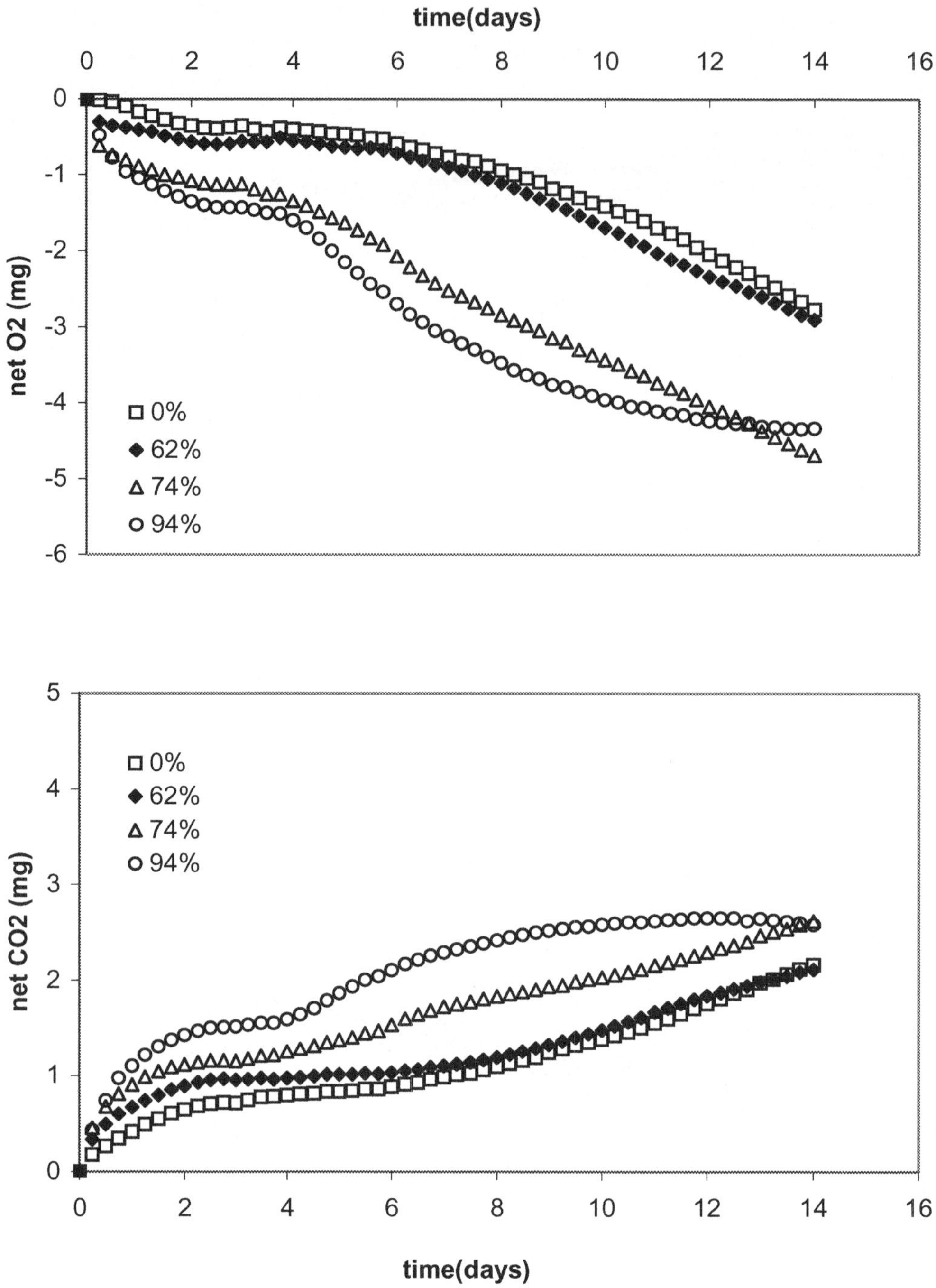

Figure 5-7. Net Oxygen Decline and Carbon Dioxide Production with Acclimated Culture of Fenton-Treated MTBE.

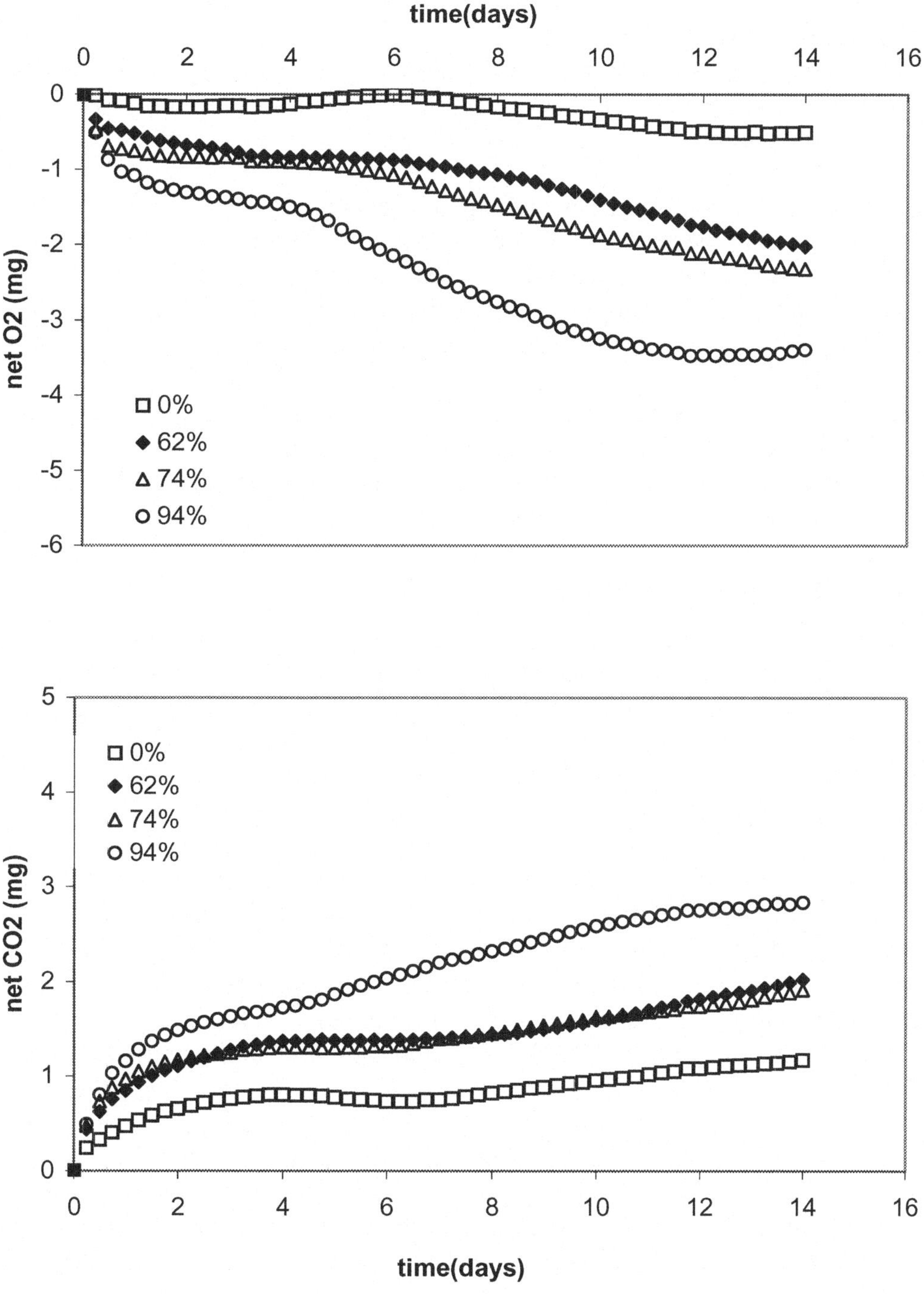

Figure 5-8. Net Oxygen Decline and Carbon Dioxide Production with Non-Acclimated Culture of Fenton-Treated MTBE.

5.2.3 UV/H_2O_2 Treated MTBE

The net decline in oxygen concentration and net production of carbon dioxide obtained by subtracting the values for endogenous respiration are plotted in Figures 5-9 and 5-10 for acclimated and non-acclimated cultures.

The results indicate that non-oxidized MTBE again biodegraded readily, though at faster rates with acclimated culture than the non-acclimated culture. In addition, pre-oxidation of MTBE at 54% and 78% levels did not produce much enhancement in biodegradation compared to non-oxidized samples. However, the sample that was pre-oxidized at 95% level exhibited significantly higher oxygen decline and carbon dioxide production.

The pH of the samples was measured at the start and end of the respirometric experiment. The result shows that there was an increase in the value of the pH in all samples. The pH increase in samples that contain non-oxidized MTBE and UV/H_2O_2-treated MTBE solutions were higher than the pH increase in the endogenous sample.

5.2.4 Comparison of Treatments Methods

The net decline in oxygen and net production of carbon dioxide over the entire test period of 14 days are presented for ozone, Fenton, and UV/H_2O_2 oxidation processes and with acclimated and non-acclimated cultures in Figure 5-11 for 75% oxidation level (80% for ozone, 74% for Fenton, and 78% for UV/H_2O_2) and in Figure 5-12 for 100% oxidation level (97% for ozone, 94% for Fenton, and 95% for UV/H_2O_2). Again, it should be noted that 75% and 100% oxidation levels were the target values, which were not always achieved.

For both oxidation levels and for the three oxidation methods, the acclimated culture exhibited a significantly higher respiration rate than the non-acclimated culture. It should be noted that the difference between the behavior of the acclimated and non-acclimated cultures during the short-term tests was not this obvious, although the μ_{max} values were slightly higher for the acclimated cultures. In addition, the results presented in Figures 5-11 and 5-12 indicate significantly higher respiration rates for the UV/H_2O_2-treated sample than those treated by ozone or Fenton reagent.

Comparison of Figures 5-11 and 5-12 also shows that the MTBE samples oxidized at 100% level consumed much higher net oxygen and produced much higher net carbon dioxide levels than those oxidized at 75% level. This indicates that the oxidation compounds produced at the 100% oxidation level were more biodegradable than those produced at 75% oxidation level.

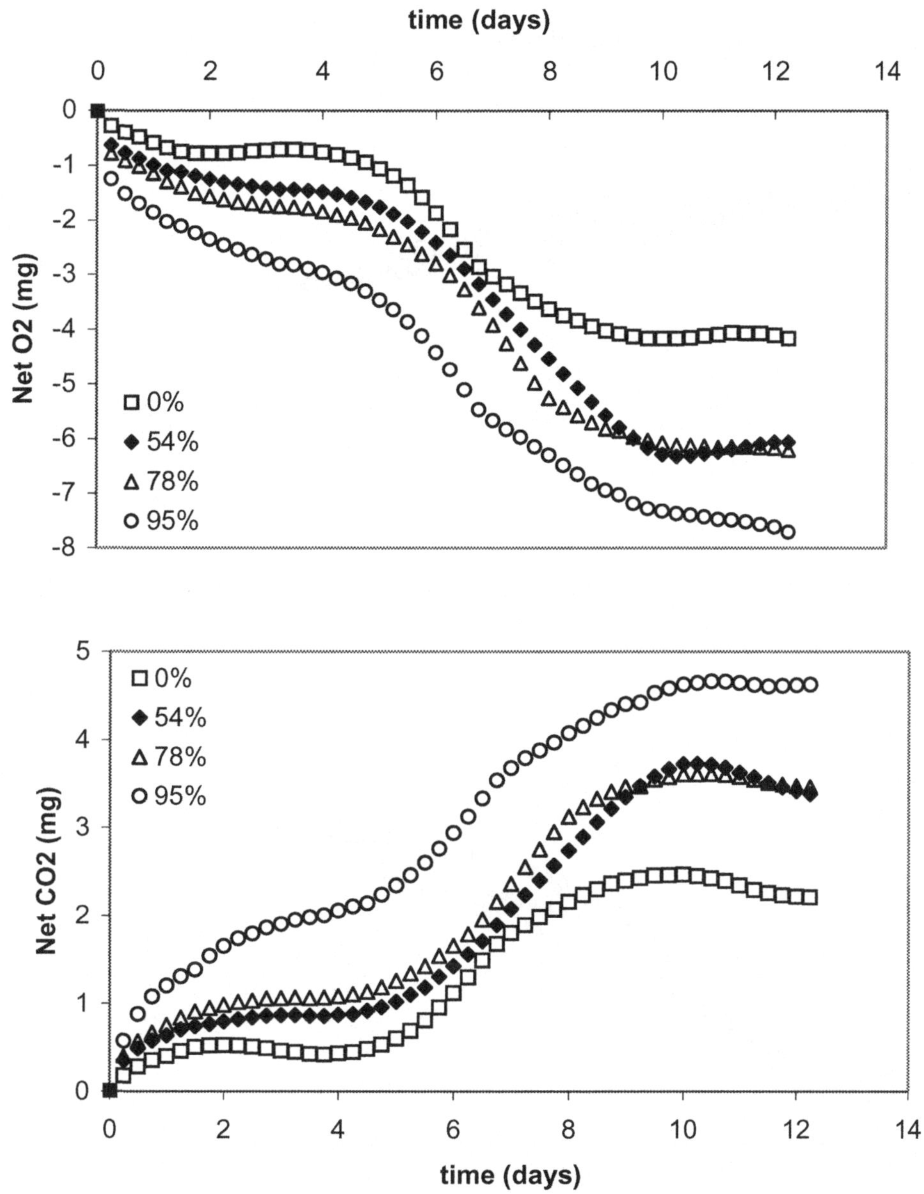

Figure 5-9. Net Oxygen Decline and Carbon Dioxide Production with Acclimated Culture of UV/H_2O_2-Treated MTBE.

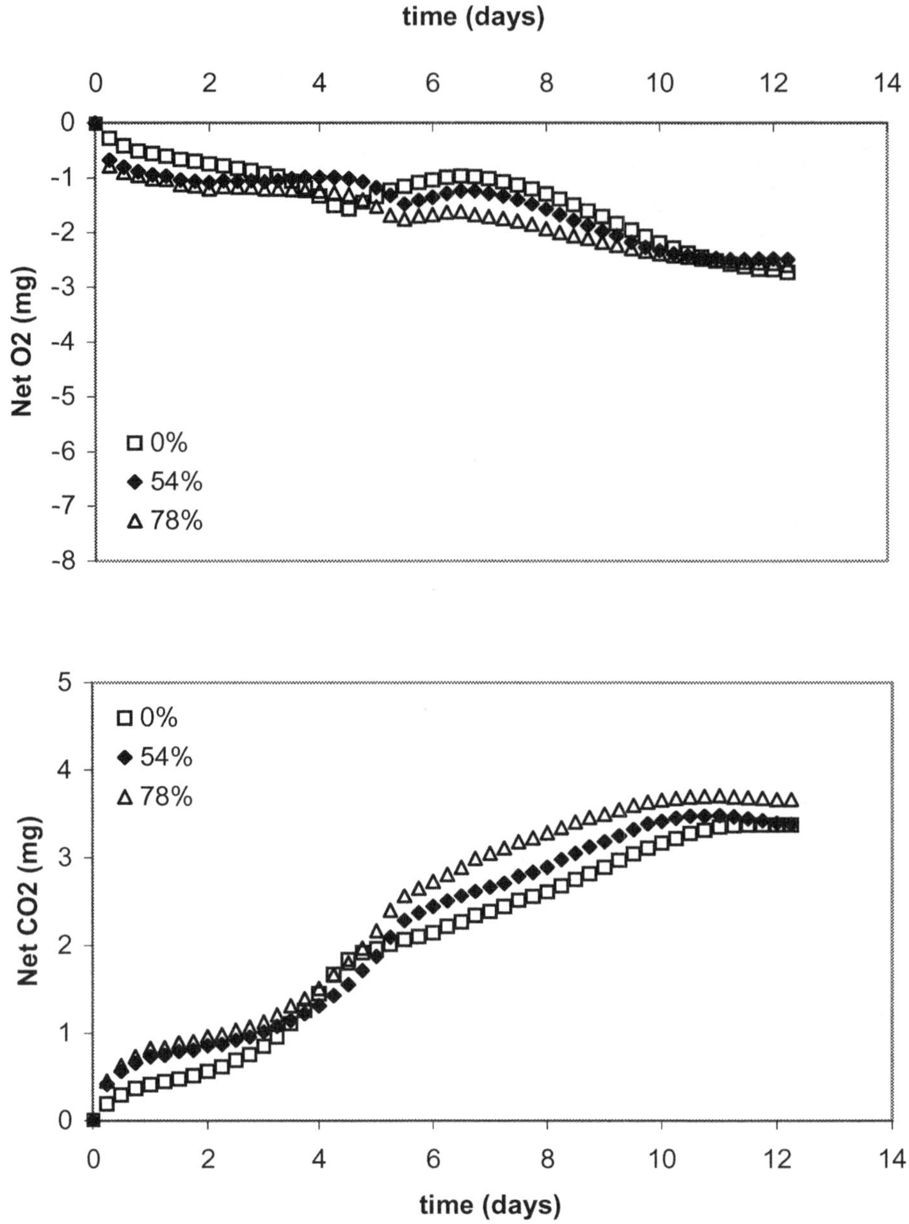

Figure 5-10. Net Oxygen Decline and Carbon Dioxide Production with Non-Acclimated Culture of UV/H_2O_2-Treated MTBE.

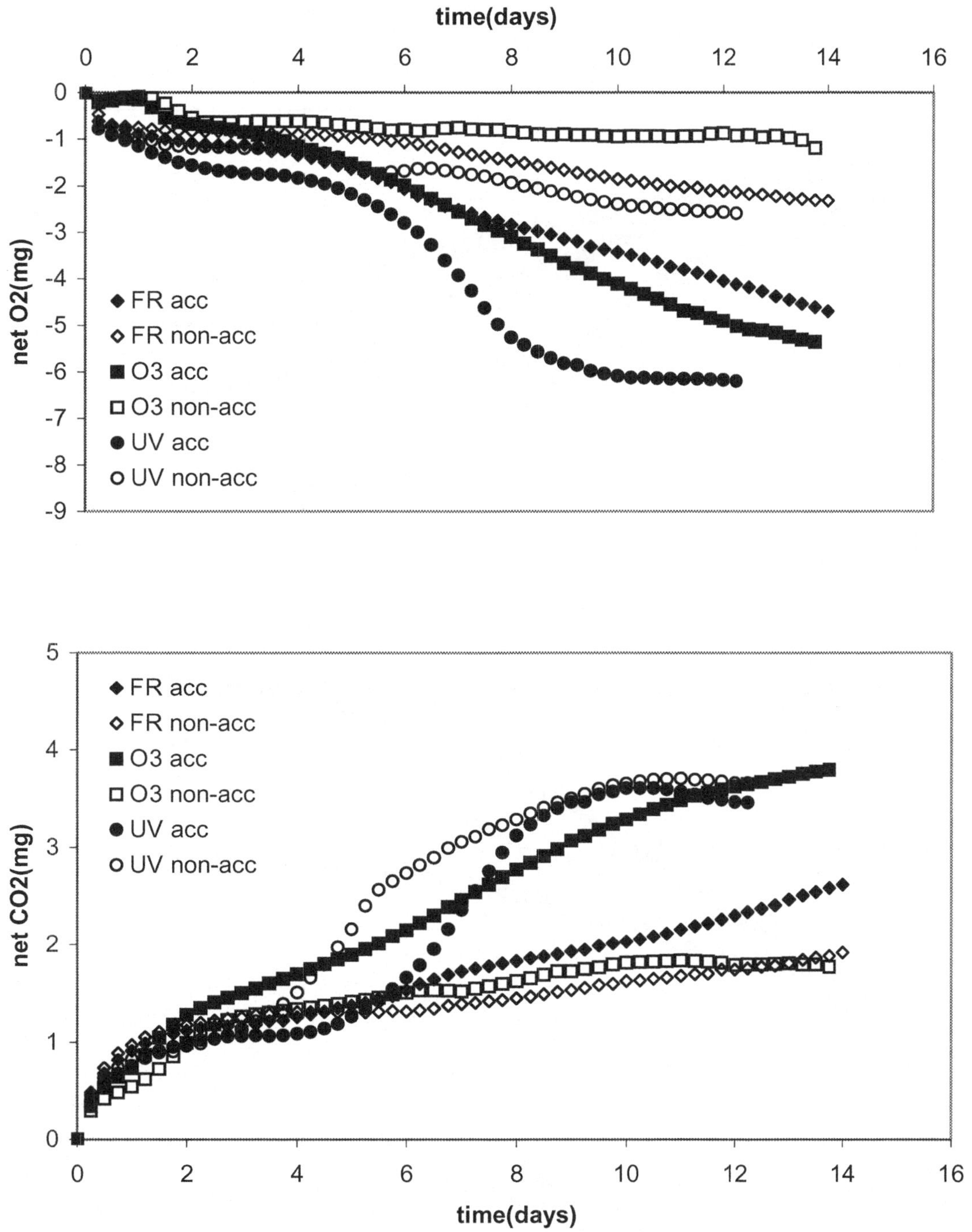

Figure 5-11. Comparison of Net O_2 Decline and CO_2 Production of Ozone, Fenton, and UV/H_2O_2-Treated Samples Using Acclimated and Non-Acclimated Cultures at 75% Oxidation Level.

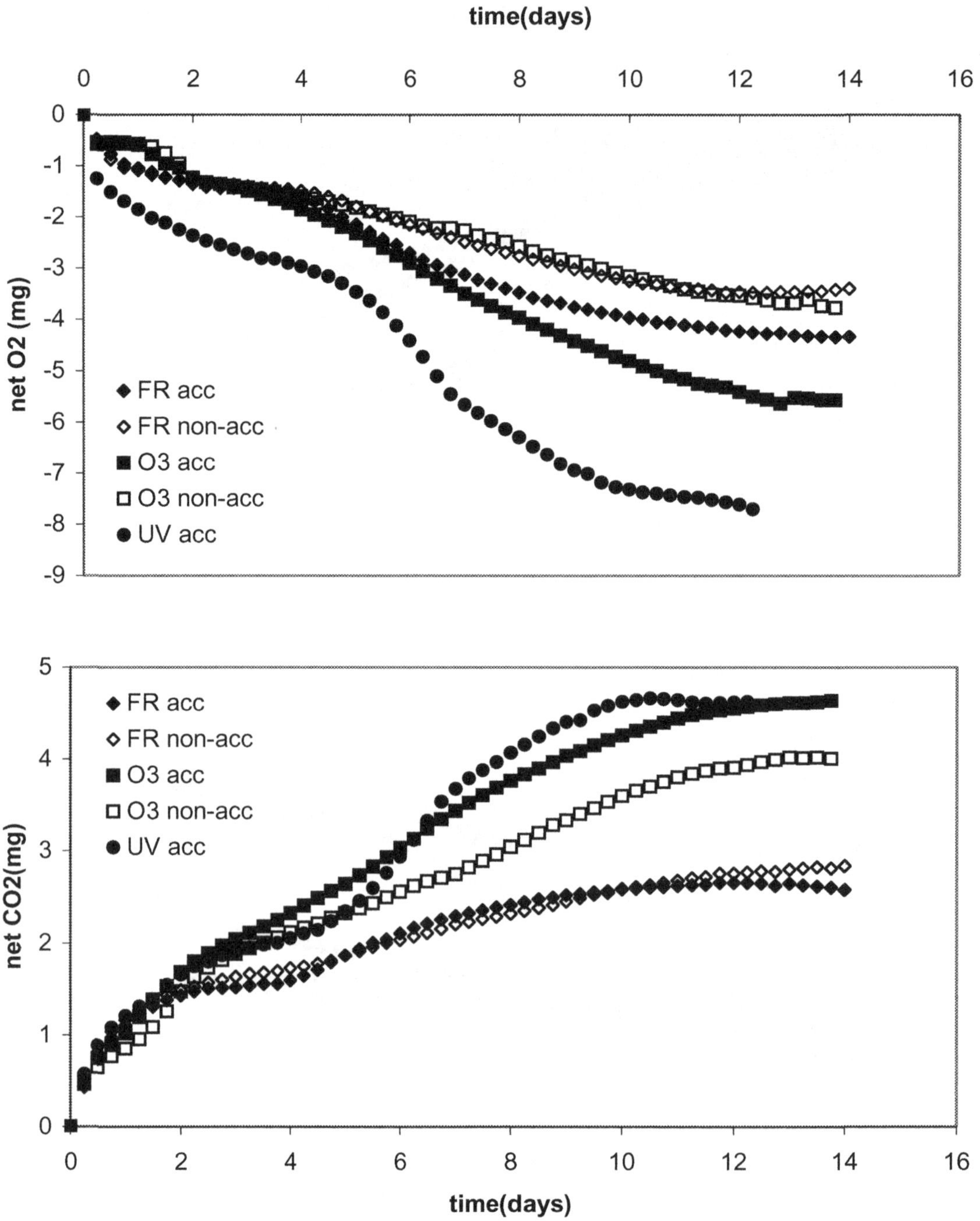

Figure 5-12. Comparison of Net O_2 Decline and CO_2 Production of Fenton, Ozone, and UV/H_2O_2-Treated Samples Using Acclimated and Non-Acclimated Cultures at 100% Oxidation Level.

5.3 Mid-Term Biodegradation Tests on UV/H_2O_2-Treated MTBE

The mid-term experiments were applied to UV/H_2O_2–treated MTBE only. The oxidation was carried at 98% level using acclimated and non-acclimated cultures. Because MTBE is easily strippable, lower oxidation levels were not checked

The results are presented in Tables 5-1 and 5.2 as average values of duplicate tests for acclimated and non-acclimated cultures, respectively. Since MTBE was practically not present in the oxidized samples, all the COD and TOC values were assumed to be contributed by the non-parent compounds. The Δ_{24}, which is the overall change in COD and TOC over the 24-hour period is also listed in each table.

Table 5-1. Biodegradation of 98% Oxidized MTBE by UV/H_2O_2 Acclimated Culture.
(The results are in non-parent COD and TOC)

Time (hr)	COD (mg/L)	TOC (mg/L)
0	228	61.2
0.5	186	48.6
1	177	45.3
1.5	160	42.9
3	144	36.6
4.5	131	33.9
6	123	31.6
21	108	25.9
24	104	26.2
Δ_{24}	124	35.0

Table 5-2. Biodegradation of 98% Oxidized MTBE by UV/H_2O_2 Non-Acclimated Culture.
(The results are in non-parent COD and TOC)

Time (hr)	$\Delta COD_{,1}$ (mg/L)	$\Delta TOC_{,1}$ (mg/L)
0	223	61.3
0.5	219	51.4
1	181	49.1
1.5	170	46.1
3	152	41.7
4.5	140	40.4
6	136	33.4
21	121	27.1
24	111	26.7
Δ_{24}	112	34.6

The results presented in the tables indicate that about 26% of the initial TOC and 22% of the initial COD contributed by 200 mg/L of MTBE were removed by biodegradation over the 24-hour period. There was no significant difference between the acclimated and non-acclimated organisms in terms of TOC and COD reduction over the test period.

A control experiment was conducted with oxidized MTBE solutions to evaluate the extent of the stripping of oxidation by-products during the 24-hour mid-term tests due to relatively vigorous aeration. This test was needed to understand whether the removal of organic matter was primarily due to biodegradation, or that air stripping also contributed to the removal of the MTBE oxidation products. The test was performed under exactly the same conditions as others, except without microorganisms present. The results are presented in Figure 5-13.

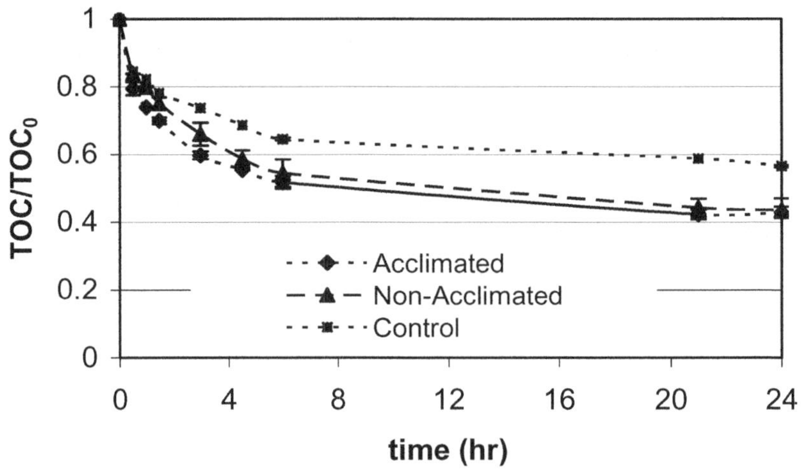

Figure 5-13. Comparison of Samples and Control Experiments for TOC Removal.

Obviously considerable amounts of by-products were removed due to stripping. However, it is well-known in the literature that the presence of particulates, in this case flocs of microorganisms in solutions, can dramatically reduce mass-transfer coefficients, i.e., the rate of stripping of organic compounds. In addition, the stripped portion of the organic matter could also be easily biodegradable if the microorganisms are present. In order to check the TOC reduction due to biodegradation only, a 50-mL sample was put in a 300-mL respirometer test bottle with microorganisms, and the respirometer was run for one day only to eliminate the "refreshing" or air exchange with surroundings in the respirometer. The TOC values of the sample were measured as 71 and 34 mg/L at the beginning and end of the experiment, respectively. The difference was 37 mg/L of TOC, and this result is very close to the TOC difference obtained by the mid-term experiments. Therefore, we have concluded that a significant portion of the COD and TOC removals shown in Tables 5.1 and 5.2 are in fact due to biodegradation of the by-products of oxidation. Numerically speaking, about 57% of the oxidation by-products of MTBE was biodegradable under the experimental conditions and when initial concentration of organic matter after the oxidation was about 60 mg/L of TOC.

CHAPTER 6.0

TOXICITY ASSESSMENT

6.1 Tests with DCDE

As explained in Section 2.2.5, the inhibition or the toxicity assessment was performed by measuring the exogenous respiration rate of microorganisms using acetate as the substrate in the presence and absence of the test chemicals. Percent exogenous inhibition was found by comparing the respiration rates. First, the endogenous respiration rate was subtracted from the exogenous respiration rate to obtain net exogenous respiration rate. Then, the inhibited endogenous rate was subtracted from the inhibited exogenous respiration rate to obtain the net inhibited exogenous respiration rate. Finally, these two values were compared by subtracting the net inhibited exogenous rate from the net exogenous respiration rate, and by dividing the difference by the net exogenous respiration rate and multiplying by 100. This gave the percent of exogenous inhibition, i.e., the degree to which the test solution inhibited the microorganisms from biodegrading acetate. All the tests were conducted in triplicate, and in the presence of two sets of controls.

Figure 6-1 shows the inhibition measured when exposing acclimated and non-acclimated activated sludge cultures to various concentrations of DCDE. Over the concentration range studied, DCDE produced some inhibition at concentrations higher than 50 mg/L. Although the extent of inhibition is small, DCDE may still not be accepted into a biotreatment system because it is not biodegradable.

6.1.1 Ozone-Treated DCDE

Samples containing 200 mg/L DCDE underwent ozonation to remove DCDE by 25%, 50%, 75%, 100%, 150%, and 200%. For 150% and 200% oxidation, the ozonation time was increased respectively by 50% and 100% of the time needed for 100% oxidation. For the toxicity tests, all the test solutions were performed in triplicate. Figure 6-2 presents the exogenous inhibition results of studies using ozonated DCDE as the test solutions.

Ozonated samples caused significant inhibition of the activities of both acclimated and non-acclimated activated sludge. Cultures that were given acetate together with the test solution were much less active than those cultures fed only acetate. Inhibition ranged from 36% to 55% for 25% to 200% oxidation of DCDE. However, samples oxidized beyond 100% showed some decrease in inhibition.

6.1.2 Fenton-Treated DCDE

Samples containing 200 mg/L DCDE underwent oxidation by Fenton reagent to remove DCDE by 25%, 50%, 75%, 100%, 150%, and 200%. The chemical addition for 150% and 200% oxidation was increased respectively by 50% and 100% of what was needed for 100% oxidation. Figure 6-3 presents the exogenous inhibition results of studies using Fenton-treated DCDE as a test solution.

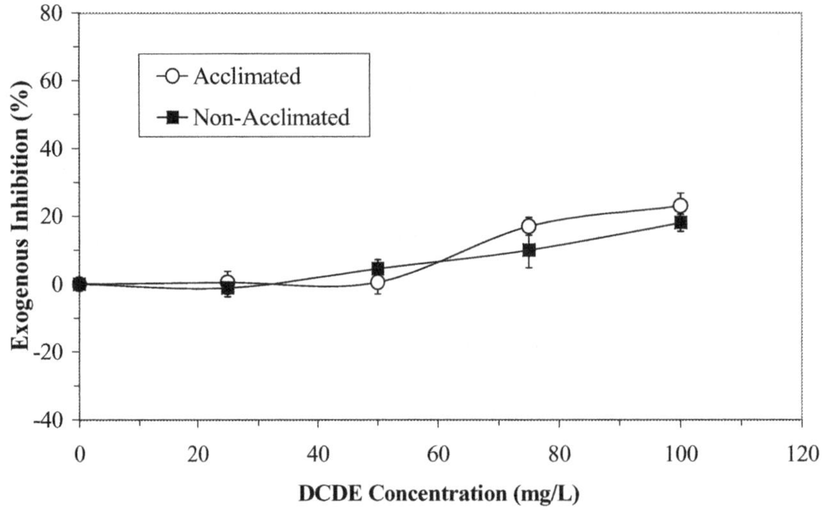

Figure 6-1. Exogenous Inhibition of Cultures in the Presence of DCDE.

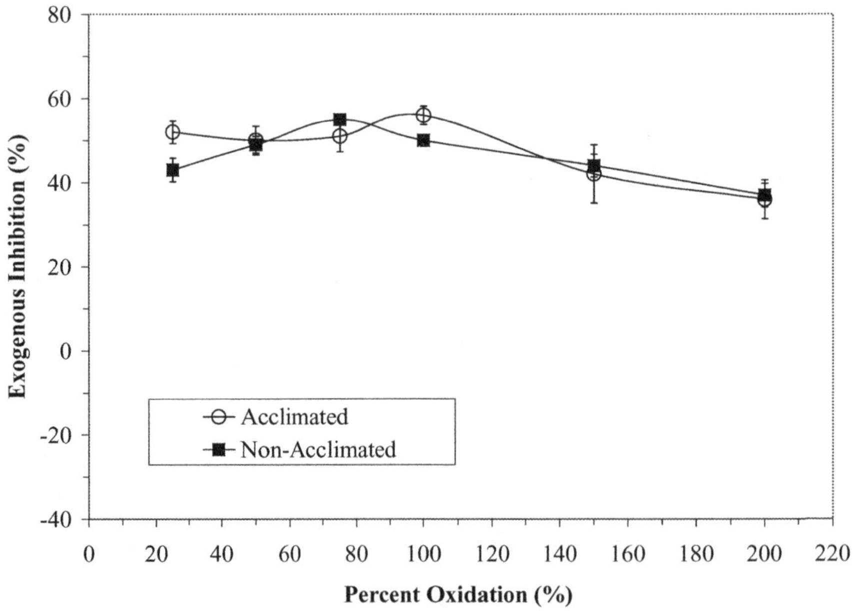

Figure 6-2. Exogenous Inhibition of Cultures Exposed to DCDE and its Oxidation By-Products After Ozonation.

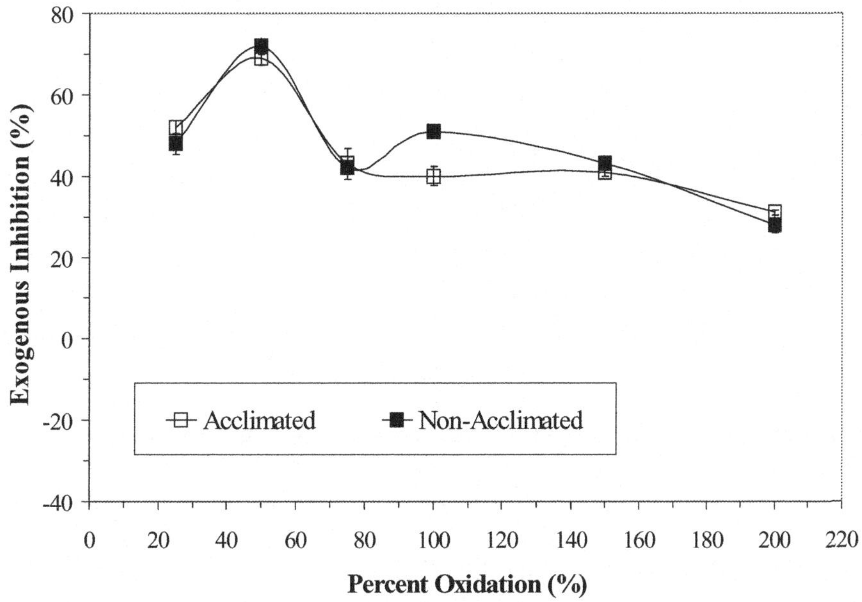

Figure 6-3. Exogenous Inhibition of Cultures Exposed to DCDE and Its Oxidation By-Products After Fenton Treatment.

Like ozonation, the test solution treated by Fenton reagent acted to decrease the respirometric activity and hence the oxygen consumption in the activated sludge. The inhibition ranged from 28% to 72% for 25% to 200% oxidation of DCDE. A decrease in inhibition was noted with extensive oxidation. After 200% oxidation, inhibition was down to 30%. However, sludge production as iron precipitates at this level of oxidation was very high. Potential inhibition from iron or catalase was checked. In the absence of Fenton-treated DCDE, the respiration rates were not inhibited by iron or catalase.

6.1.3 UV/H_2O_2-Treated DCDE

Figure 6-4 present the exogenous inhibition results of studies using UV/H_2O_2-treated DCDE as a test solution.

The results were very favorable for solutions oxidized with UV/H_2O_2. Not only was there little or no detectable inhibition, but the oxidized solution served as an excellent secondary substrate. Samples that contained both acetate and the test solution showed an increase in oxygen consumption over the samples with acetate and no test solution. The negative inhibition values indicated that not only was there no detectable inhibition but that some of the DCDE by-products were actually readily biodegradable.

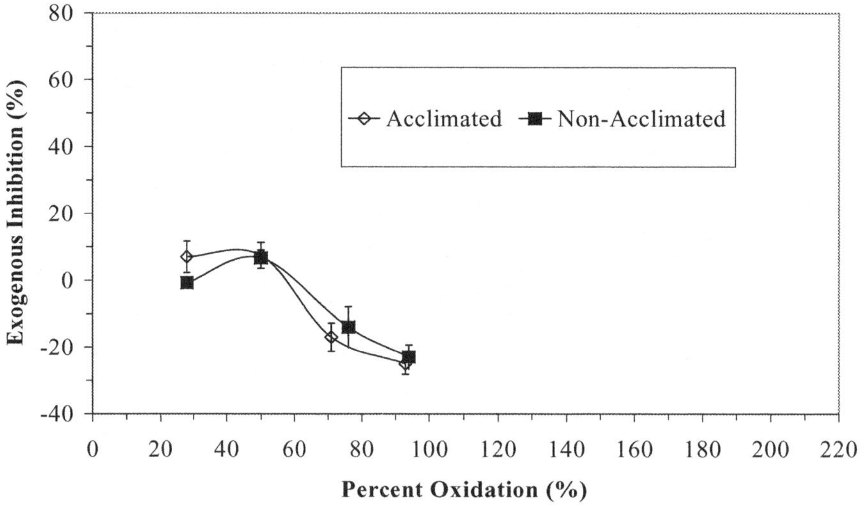

Figure 6-4. Exogenous Inhibition of Cultures Exposed to DCDE and Its Oxidation By-Products After UV/H_2O_2 Treatment.

6.1.4 Comparison of Treatments for DCDE

Figure 6-5 presents results from experiments involving acclimated activated sludge and all three chemical oxidation techniques for comparison. Figure 6-6 presents results from experiments involving non-acclimated activated sludge and all three chemical oxidation techniques for comparison.

Figure 6-7 also shows results from experiments involving acclimated activated sludge cultures and all three chemical oxidation techniques as well as untreated DCDE. In this figure, the exogenous inhibition is shown as related to DCDE concentration. Figure 6-8 presents the results with non-acclimated cultures.

Acclimation of activated sludge did not prove to be beneficial in this study. Furthermore, solutions of DCDE tested after ozonation and modified Fenton treatment showed a similar range of inhibition. Solutions tested after UV/H_2O_2 treatment caused minimal exogenous inhibition after 25% and 50% oxidation and no detectable inhibition for higher oxidation levels. The differences in inhibition among the oxidation methods were likely due to the different types and concentrations of the by-products of the chemical oxidation processes. However, the inhibition studies indicate the UV/H_2O_2 treatment is the preferred oxidation method.

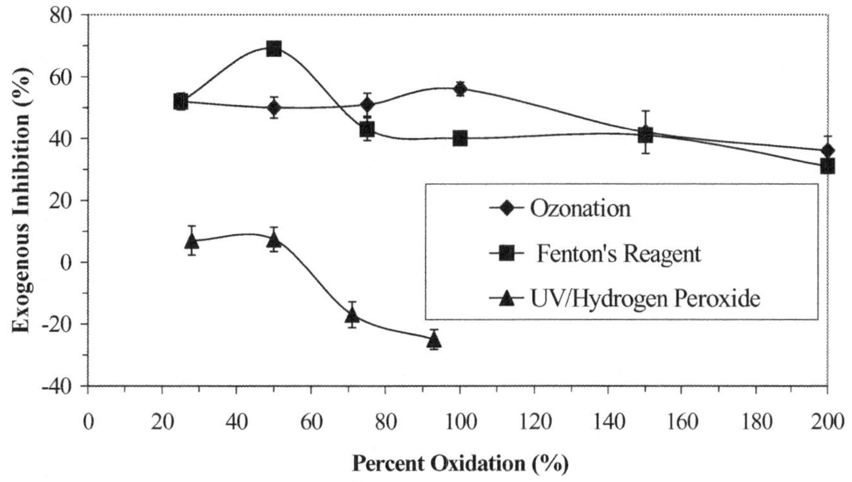

Figure 6-5. Comparison of Inhibition on Acclimated Culture of Oxidized DCDE.

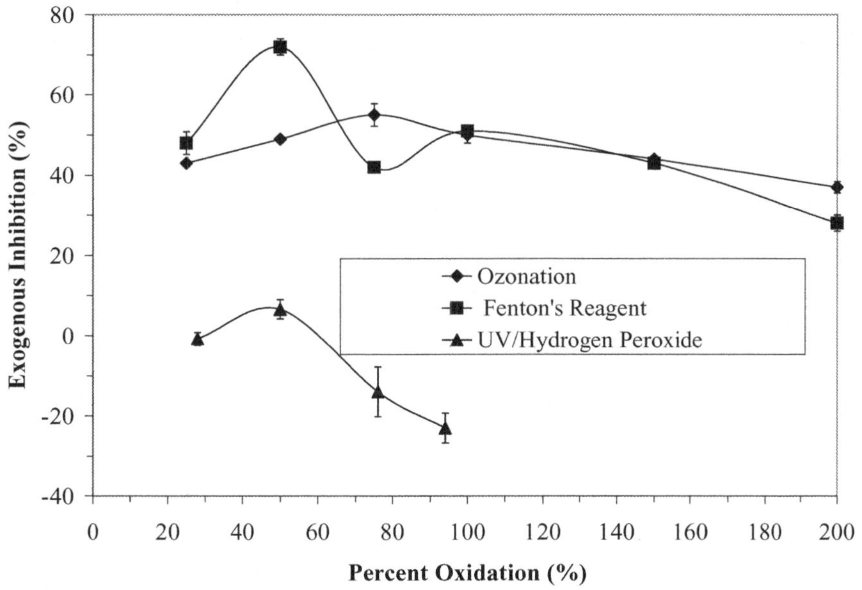

Figure 6-6. Comparison of Inhibition on Non-Acclimated Culture of Oxidized DCDE.

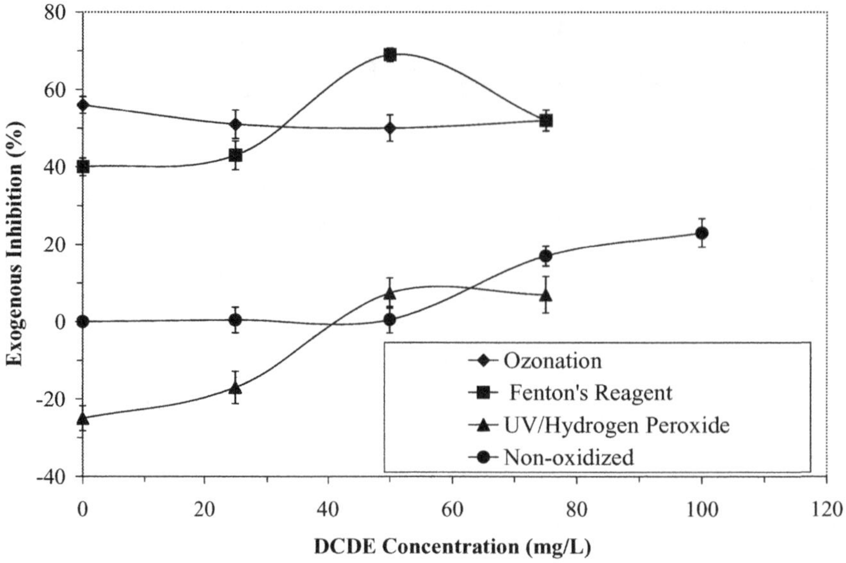

Figure 6-7. Comparison of Inhibition on Acclimated Cultures for Oxidized and Non-Oxidized DCDE.

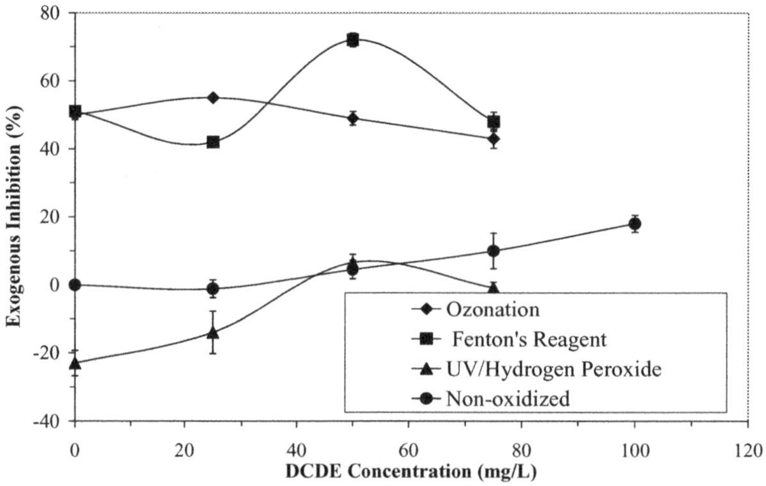

Figure 6-8. Comparison of Inhibition on Non-Acclimated Cultures for Oxidized and Non-Oxidized DCDE.

6.2 Tests with MTBE

The results of inhibition tests conducted by non-oxidized MTBE are presented in Tables 6-1 and 6-2 for acclimated and non-acclimated cultures, respectively.

Table 6-1. Inhibition of Respiration of Acclimated Culture Due to Non-Oxidized MTBE.

MTBE Concentration (mg/L)	Percent Inhibition (%)	Standard Deviation (%)
0	0.0	0.0
25	8.8	2.3
50	1.5	3.3
75	-1.9	3.2
100	3.5	4.8

Table 6-2. Inhibition of Respiration of Non-Acclimated Culture Due to Non-Oxidized MTBE.

MTBE Concentration (mg/L)	Percent Inhibition (%)	Standard Deviation (%)
0	0.0	0.0
25	9.7	2.0
50	1.1	1.5
75	0.38	0.7
100	1.7	0.3

It is obvious that MTBE did not cause any inhibition on the acclimated and non-acclimated cultures.

6.2.1 Ozone-Treated MTBE

The results of inhibition tests on ozone-treated MTBE are presented in Tables 6-3 and 6-4 for acclimated and non-acclimated cultures.

Table 6-3. Inhibition of Respiration of Acclimated Culture Due to Ozonated MTBE.

Percent Oxidation (%)	MTBE Concentration (mg/L)	Percent Inhibition (%)	Standard Deviation (%)
196	0	-67	8.4
147	0.15	-49	1.4
98	2.0	-58	1.3
83	16.5	-112	0.13
50	50	-76	8.3
44%	51	-27	1.0

Table 6-4. Inhibition of Respiration of Non-Acclimated Culture Due to Ozonated MTBE.

Percent Oxidation (%)	MTBE Concentration (mg/L)	Percent Inhibition (%)	Standard Deviation (%)
196	0	-66	7.6
147	0.15	-47	9.6
98	2.0	-29	0.73
83	17	-95	1.6
50	50	-81	4.7
44	51	-27	3.4

Ozonation by-products of MTBE did not cause any inhibition. In fact, the by-products served as an excellent secondary substrate, enhancing the respirometric activity substantially. MTBE oxidized by about 80% showed the highest negative inhibition, indicating that the by-products at that oxidation level were probably more biodegradable. The results are not very different between the acclimated and the non-acclimated sludge.

6.2.2 Fenton-Treated MTBE

The results of inhibition tests on Fenton-treated MTBE are presented in Tables 6-5 and 6-6 for acclimated and non-acclimated cultures.

Table 6-5. Inhibition of Respiration of Acclimated Culture Due to Fenton-Treated MTBE.

Percent Oxidation (%)	MTBE Concentration (mg/L)	Percent Inhibition (%)	Standard Deviation (%)
143	4.0	-81	12
95	5.0	-11	1.1
75	25	-104	6.5
50	50	-28	2.7
28	71	+10.2	0.4

Table 6-6. Inhibition of Respiration of Non-Acclimated Culture Due to Fenton-Treated MTBE.

Percent Oxidation (%)	MTBE Concentration (mg/L)	Percent Inhibition (%)	Standard Deviation (%)
143	4.0	-103	9.7
95	5.0	-15	1.3
75	25	-91	1.9
50	50	-29	2.7
28	71	+11.3	2.6

The results obtained for Fenton-treated MTBE are similar to those observed for ozonated MTBE.

6.2.3 UV/H_2O_2 -Treated MTBE

The results of inhibition tests on UV/H_2O_2-treated MTBE are presented in Tables 6-7 and 6-8 for acclimated and non-acclimated cultures.

Table 6-7. Inhibition of Respiration of Acclimated Culture Due to UV/H_2O_2 -Treated MTBE.

Percent Oxidation (%)	MTBE Concentration (mg/L)	Percent Inhibition (%)	Standard Deviation (%)
92	8.3	-30	3.1
68	32	-25	8.8
50	50	-34	5.4
28	69	-8.3	3.4

Table 6-8. Inhibition of Respiration of Non-Acclimated Culture Due to UV/H_2O_2 -Treated MTBE.

Percent Oxidation (%)	MTBE Concentration (mg/L)	Percent Inhibition (%)	Standard Deviation (%)
92	8.3	-38	8.5
68	32	-28	2.4
50	50	-39	2.6
28	69	-26	8.2

6.2.4 Comparison of Treatments for MTBE

The exogenous inhibition results obtained for three different oxidation processes are plotted in Figure 6-9 as a function of percent oxidation of MTBE. The average of the values obtained for the acclimated and the non-acclimated cultures were used for this plot.

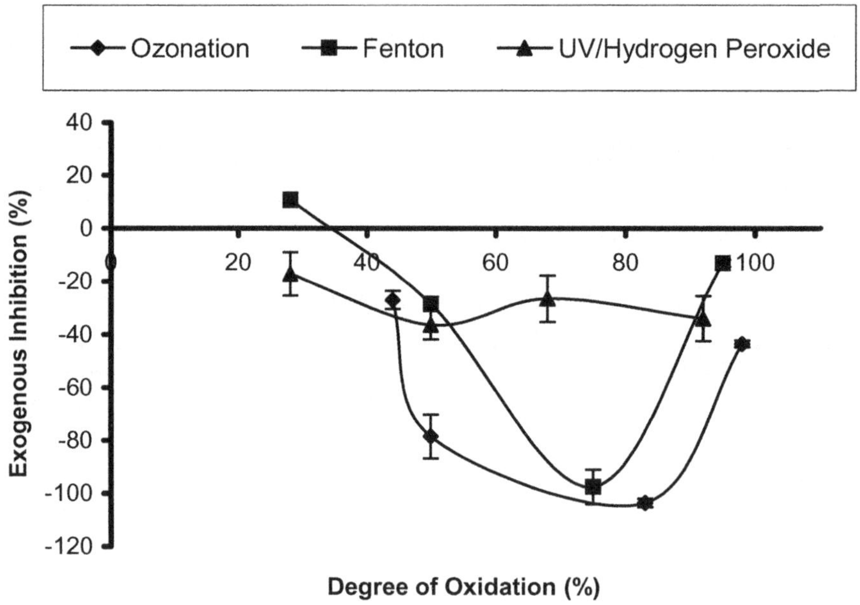

Figure 6-9. Comparison of Percent Inhibition for Oxidized MTBE by the Three Oxidation Methods.

From the data presented above, the following can be concluded:

- Acclimated and non-acclimated cultures did not show a significant difference in their reactions to MTBE and its by-products of oxidation.
- For the most part, MTBE and its by-products of oxidation caused no exogenous inhibition (negative inhibition). In fact, the by-products served as an excellent secondary substrate, enhancing activity substantially. This implies that in a full-scale reactor, there would be no concern for disrupting an activated sludge system.
- Activated sludge showed less overall enhanced activity after UV/H_2O_2 oxidation compared to Fenton and ozonation.
- Different degrees of oxidation produced various peaks. Assuming that the experimental procedure was accurate and consistent and that activated sludge did not vary dramatically on a day-to-day basis, it is possible that these fluctuations were due to the build-up of different by-products, some of which were more biodegradable than the others.

CHAPTER 7.0

SEQUENCING BATCH REACTOR (SBR) TESTS

7.1 Assessment of Biodegradation

Biodegradation for oxidized DCDE was evaluated based on TOC data because this data were shown to be more reliable than the COD data (smaller standard deviation for TOC than for COD). Large variations in COD data from some DCDE experiments may have been due to instrumental error or interferences caused by oxidation by-products. Both the COD and TOC results from the experiments conducted with MTBE contained small errors, so both were used for evaluation of biodegradation of oxidized MTBE solutions.

Figure 7-1 shows a graphical example for assessment of biodegradable and nonbiodegradable portions of the feed solution that was introduced into the reactors. The nonbiodegradable portion of the feed to the control reactor was attributed to the organic compounds nonbiodegradable in the Syntho solution. In addition some microbial by-products may not degrade completely, further contributing to the nonbiodegradable fractions in the reactors.

Biodegradation of the by-products of oxidation was determined by using the following equations:

$$\%C_{non\text{-}bio} = \%C_{reduction,1} - \%C_{reduction,2}$$

$$\%C_{by\text{-}products} = (C_{by\text{-}products}/C_{total}) \times 100$$

$$\%Bio = [(\%C_{by\text{-}products} - \%C_{non\text{-}bio}) \div \%C_{by\text{-}products}] \times 100$$

where:

$\%C_{non\text{-}bio}$ = nonbiodegradable TOC contributed by oxidized solution (%)
$\%C_{reduction,1}$ = average TOC reduction in the control reactors (%)
$\%C_{reduction,2}$ = average TOC reduction in the reactors with oxidized chemical (%)
$\%C_{by\text{-}products}$ = percent TOC contributed from by-products of oxidation (%)
$C_{by\text{-}products}$ = TOC contributed by oxidized solution (mg/L)
C_{total} = total TOC of the feed solution (mg/L)
$\%Bio$ = biodegradation of the by-products of oxidation in oxidized solution (%)

These equations assume that $\%C_{non\text{-}bio}$ is contributed only by nonbiodegradable by-products of oxidation. Similar equations were used when calculating biodegradation from COD data.

Control Reactor

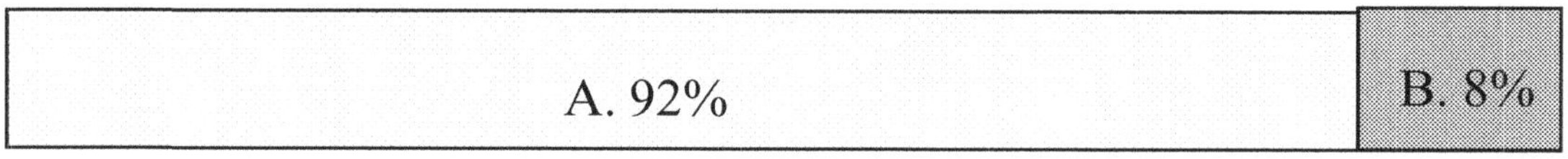

a) The biodegradable portion of Syntho and supplement solution (if used) is represented by Section A, whereas the nonbiodegradable portion of Syntho and by-products of microbial degradation is represented by Section B.

Reactor with Oxidized Chemical

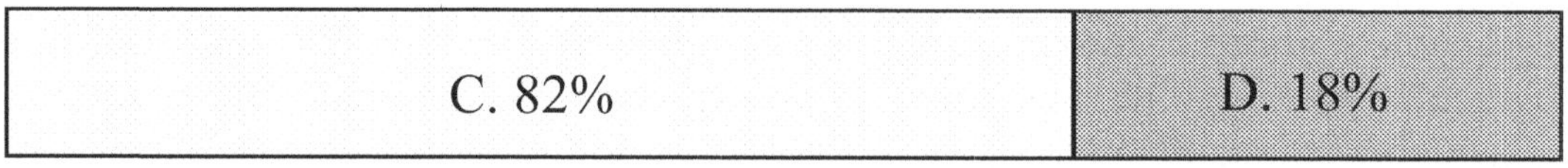

b) The biodegradable portion of Syntho, supplement solution (if used), and oxidized by-products is represented by Section C, whereas the nonbiodegradable portion of Syntho, by-products of oxidation and microbial degradation is represented by Section D.

Reactor with Oxidized Chemical

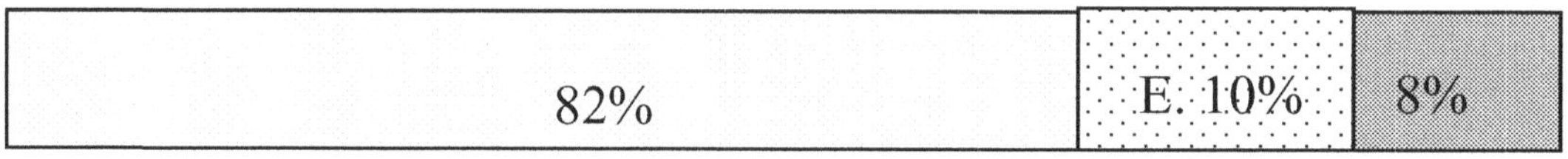

c) The nonbiodegradable portion of the by-products of oxidation, which is represented by Section E, is estimated by subtracting Section B (the nonbiodegradable fraction of the control reactor) from Section D (the nonbiodegradable fraction of the reactor with oxidized chemical).

Figure 7-1. A Graphic Example of Estimation of the Biodegradable and Nonbiodegradable Fractions of the Feed Solutions.

7.2. Tests with DCDE

The fraction of COD and TOC contributed by the oxidized chemicals to the feed solutions was chosen at various levels to determine the effects of varying amounts of oxidized chemicals on biomass population. The remaining fraction of the feed was contributed by Syntho and distilled water. Six reactors were used for each experiment. The reactors were labeled as Reactor 1 (R1) through Reactor 6 (R6). The initial experiments were conducted with an influent COD value of ~1,100 to 1,500 mg/L and were referred to as "concentrated feed" experiments. This level of COD was mainly attained by the use of a supplement solution. Later in the research, it was decided to lower the COD concentration to ~300 mg/L to simulate the incoming COD of a conventional wastewater treatment plant. These experiments were referred to as "dilute feed".

In the concentrated experiments, the amount of COD and TOC contributed by oxidized chemicals was not as significant as the COD and TOC contributed by the supplement solution. In the dilute experiments, the feed contained a higher percentage of COD and TOC contributed by oxidized chemical. Table 7-1 shows the target contributions of DCDE by-products to COD and TOC for each experiment, as well as other target parameters for each experiment.

Table 7-1. Target COD and TOC Contributions of Oxidized DCDE to the Feed Solution.

Experiment No.	Chemical	Oxidative Treatment	Total COD Input	Waste Volume	Oxidation Percentage	TOC Contribution	COD Contribution
7	DCDE	Ozone	1500	75%	75%	9%	8%
8	DCDE	Ozone	300	40%	100%	32%	29%
9*	DCDE	Ozone	300	40%	100%	32%	29%
3	DCDE	Fenton	1500	25%	75%	2%	2%
4	DCDE	Fenton	1500	25%	100%	2%	3%
5	DCDE	Fenton	1500	75%	75%	5%	7%
6	DCDE	Fenton	1500	75%	100%	6%	10%
1	DCDE	UV/H_2O_2	1500	75%	75%	9%	8%
2	DCDE	UV/H_2O_2	300	40%	100%	32%	29%

* Experiment 9 is a repeat experiment

The values presented in Table 7-1 were calculated assuming 75% and 100% oxidations, TOC values of 500 and 100 mg/L for concentrated and dilute experiments, respectively, and COD values of 1500 and 300 mg/L for concentrated and dilute experiments, respectively. These calculations were made to get an estimate of by-product contribution to COD and TOC for the SBR experiments. As mentioned previously, these were target values and were not always attained.

7.2.1 Ozone-Treated DCDE

Figures 7-2 to 7-4 present the results of Experiment No. 7 with ozonated DCDE. The solids were wasted every other day if the MLSS concentration exceeded 4000 mg/L. The results presented in the figures are the average values of two data points on two simultaneously run reactors. The standard errors estimated from the data are also presented.

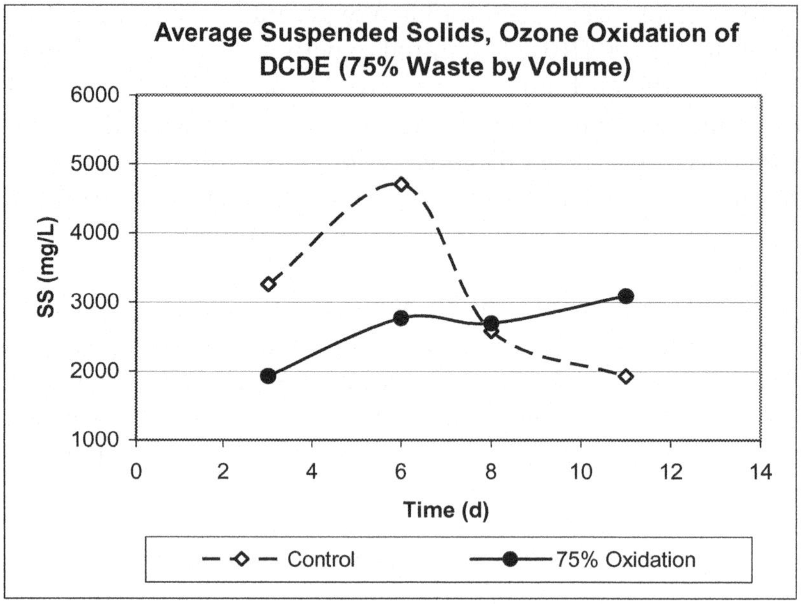

Figure 7-2. Average Suspended Solids vs. Time (Experiment 7).

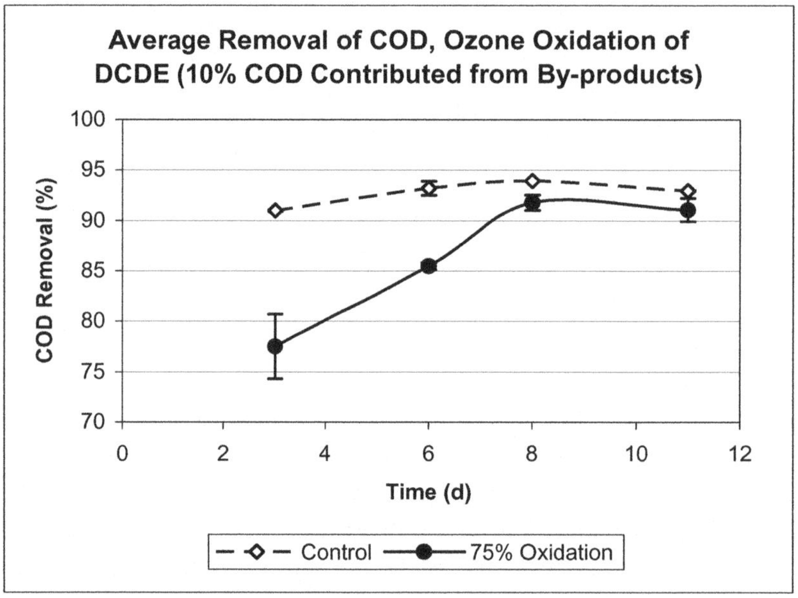

Figure 7-3. Average COD Removal Over Time (Experiment 7).

As shown in the figures, the reactors receiving oxidized chemical were about 10% less efficient in COD and TOC removal in the beginning of the experiment. However, the COD and TOC removal efficiencies increased throughout the duration of the experiment. The results of the toxicity tests showed that DCDE oxidized by ozone exhibited about 50% exogenous inhibition for all oxidation levels. However, according to the SBR experiments, the initial slight inhibition caused by ozonation by-products was reduced with time, probably because of acclimation.

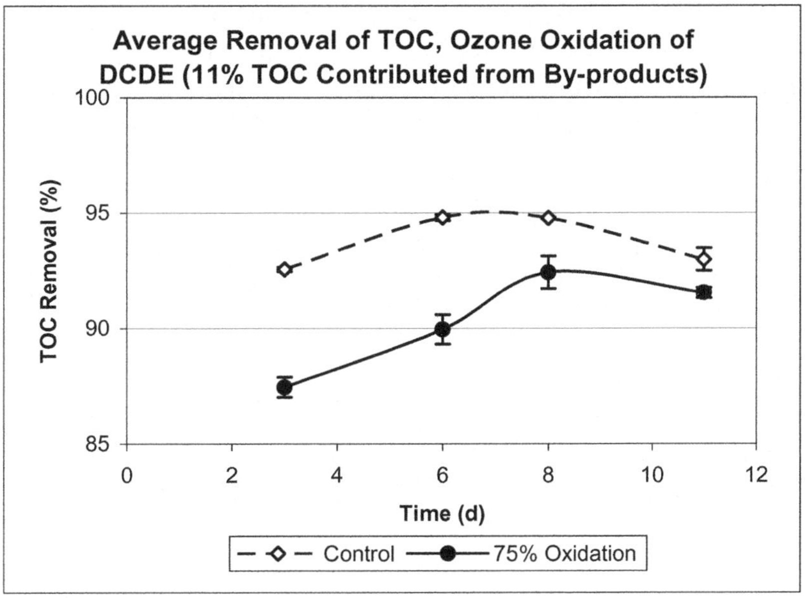

Figure 7-4. Average TOC Removal Over Time. (Experiment 7)

Estimated by-product biodegradation presented in Table 7-2 as percentages clearly shows increased amount of biodegradation with time.

Table 7-2. Estimation of Percent Biodegradation of DCDE By-Products. (Experiment 7)

By-Product Biodegradation (%)				
Day	**Average Based on COD**	**Range**	**Average Based on TOC**	**Range**
3	–	–	59	55 – 63
6	26	13 – 38	62	62
8	77	73 – 82	77	75 – 79
11	80	68 – 91	87	85 – 89

The following figures present the results of Experiment 8 on ozonated DCDE with a dilute feed stream. The MLSS concentration ranged between 1200-2000 mg/L for this experiment, as seen in Figure 7-5. The reactors containing oxidized DCDE had a lower biomass concentration than the controls. The COD data showed a large variation in reduction values from 60%–80% for the reactors receiving oxidized chemical (Figure 7-6).

Figure 7-7 shows an initial 10% difference in TOC removal efficiency between the control and the reactors receiving oxidized chemical. However, TOC removal increased by about 5% with time in the reactors receiving oxidized DCDE. Again, an initial inhibition of the by-products of DCDE seems to be followed by subsequent acclimation to the by-products. Earlier toxicity experiments revealed that DCDE oxidized by 75% and 100% by ozone caused 50% inhibition on activated sludge. However, by-products of 71% and 90% oxidation by ozone showed no toxicity in long-term experiments.

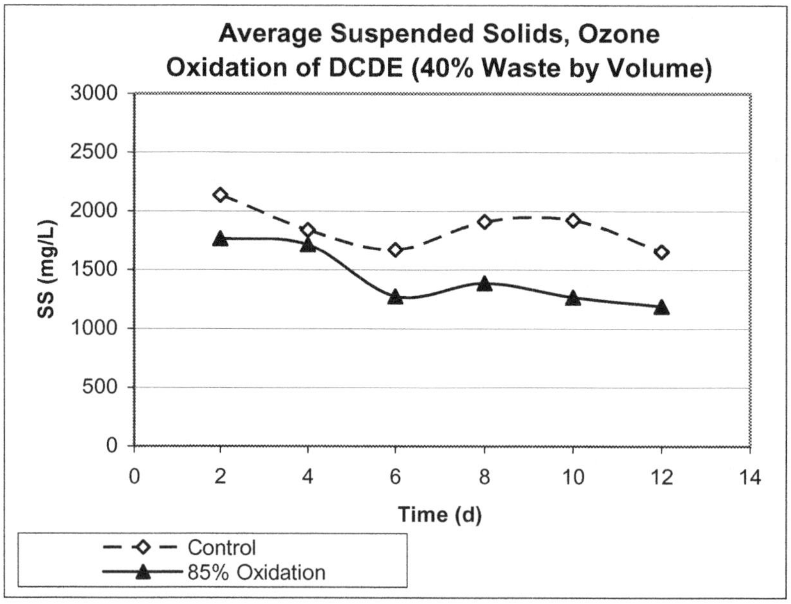

Figure 7-5. Average Suspended Solids vs. Time. (Experiment 8)

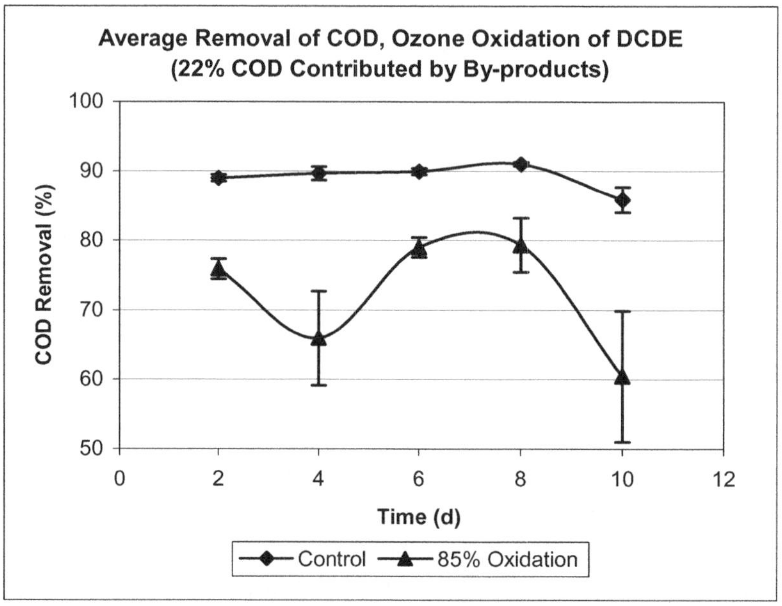

Figure 7-6. Average COD Removal vs. Time. (Experiment 8)

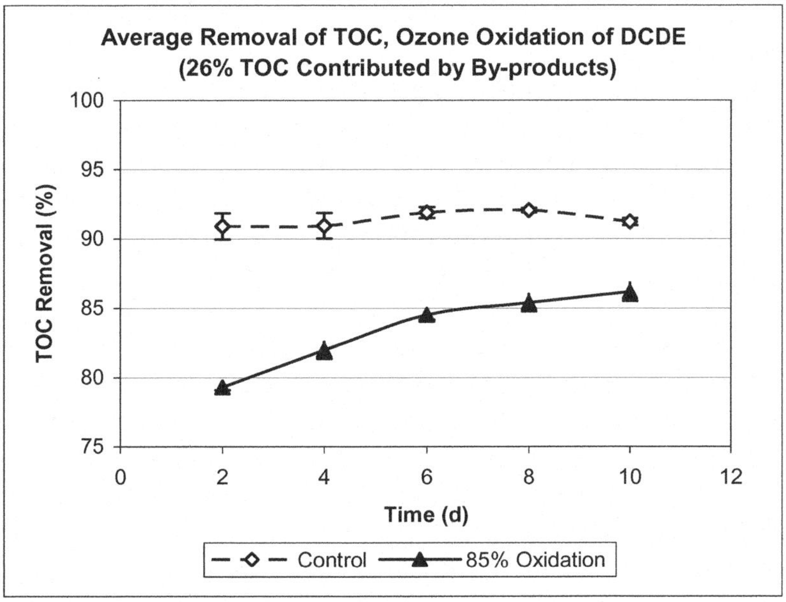

Figure 7-7. Average TOC Removal vs. Time. (Experiment 8)

Table 7-3 indicates a slight increase in percent biodegradation throughout the course of the experiment based on TOC values. In a repeat experiment (Experiment 9), utilizing DCDE oxidized by ozone, a similar trend was seen (data not shown here but presented in Christensen, 2002). The mid-term experiments showed that about 37% of DCDE by-products of ozonation (at 94% oxidation level) were biodegradable. Nevertheless, the SBR experiments indicated about 80% biodegradation. Furthermore, a significant difference between the results of the concentrated and dilute feed experiments was detected.

Table 7-3. Estimation of Percent Biodegradation of Ozonation By-Products of DCDE. (Experiment 7)

By-Product Biodegradation (%)				
Day	Average Based on COD	Range	Average Based on TOC	Range
2	48	47 – 48	56	56
4	–	–	67	66 – 69
6	65	51 – 62	69	68 – 70
8	38	30 – 46	71	69 – 75
10	–	–	78	78 – 82

7.2.2 Fenton-Treated DCDE

The results of the SBR experiments performed utilizing Fenton-treated DCDE are presented in Figures 7-8 to 7-10. All Fenton experiments were performed using concentrated feed. The target DCDE oxidation values were 75% and 100%. However, analysis of samples after oxidation indicated that only 53% and 87% of DCDE was oxidized, respectively. In these experiments, 25% by volume of the oxidized solutions were fed for the first 6 days. After day 6, the volumes of oxidized solutions were increased to 75%.

After day 6, the MLSS concentration started to steadily increase from 5,000 to about 9,000 mg/L in the reactors that received oxidized DCDE (Figure 7-8). However, the MLSS concentration in the control reactors was stable throughout the duration of the experiment. The only mechanism for solids wasting in all reactors was biomass leaving with the effluent during the decant period of each cycle. Therefore, the increase in biomass concentration in the reactors receiving oxidized solutions can be explained by the presence of dissolved iron in the Fenton-oxidized solutions, which probably facilitated coagulation, thus inducing better settling of the sludge in the reactors receiving oxidized chemical. This phenomena would reduce the amount of solids withdrawn during each decant period. It should also be mentioned that when the percentage of the oxidized solution was increased from 25% to 75% after day 6, the dissolved iron concentration most likely increased as well, causing more solids accumulation.

According to Figures 7-9 and 7-10, the reactors receiving oxidized DCDE performed similar to the controls up to day 6. The overlap in the data suggests that all reactors performed with similar efficiency. However, after the volume of oxidized chemical was increased from 25% to 75%, the COD and TOC removal efficiencies started to decline. The reactors receiving 53% oxidized chemical exhibited more stress than the reactors receiving 87% oxidized chemical. The stress could be due to introduction of toxic by-products to the reactors in large quantities. In fact, the toxicity tests on Fenton oxidation by-products showed inhibition of exogenous respiration by 70% and 40% for oxidation levels of 50% and 100%, respectively. The long-term experiments also showed that DCDE oxidized by Fenton resulted in toxicity. It should be noted that the reductions in the COD and TOC removal efficiencies occurred in spite of significant increases in the biomass concentration. This observation further strengthens the contention that the reactors were exposed to large amounts of inhibitory by-products of Fenton treatment.

As seen from Table 7-4, the average biodegradation calculated for the reactors containing 53% oxidized chemical was less than the reactors containing 87% oxidized chemical. The mid-term and long-term experiments also showed more biodegradation of the by-products with increasing percentages of oxidation. The average biodegradation obtained from mid-term experiments was about 25% for DCDE oxidized by Fenton at 90% level.

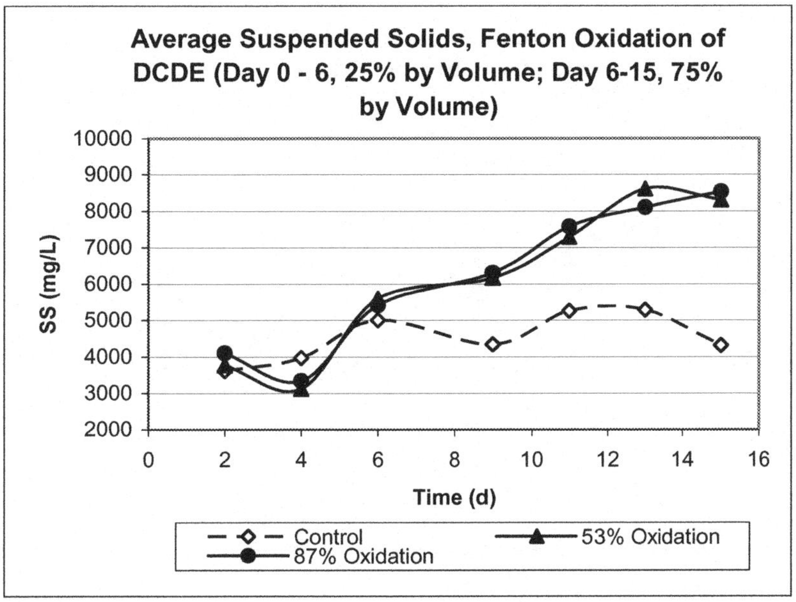

Figure 7-8. Average Suspended Solids vs. Time. (Experiments 3-6)

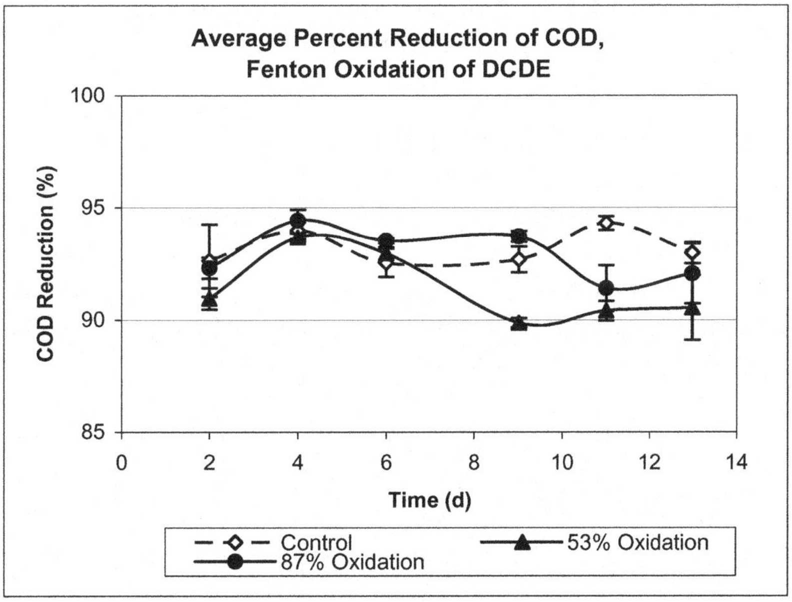

Figure 7-9. Average COD Removal Over Time. (Experiments 3-6)

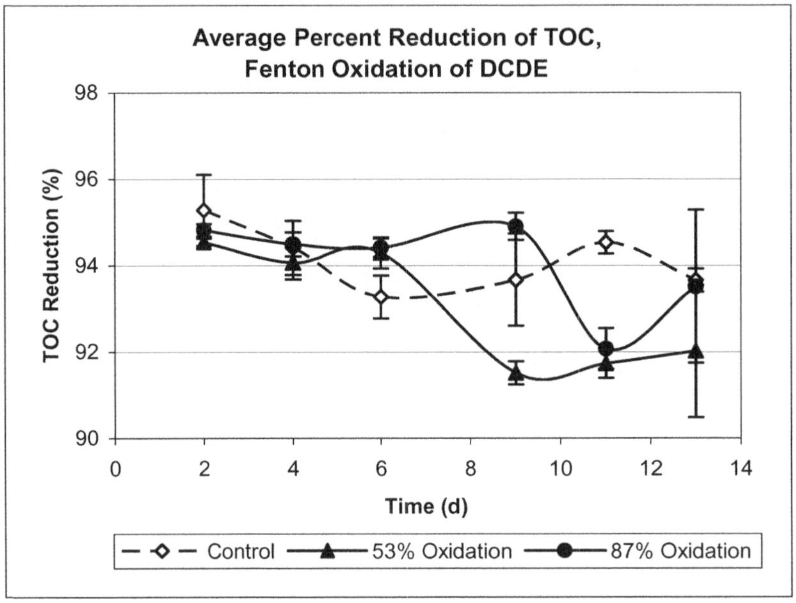

Figure 7-10. Average TOC Removal Over Time. (Experiments 3-6)

Table 7-4. Estimation of Percent Biodegradation of Fenton By-Products of DCDE. (Experiment 6)

	By-Product Biodegradation (%)							
	53% Oxidation		87% Oxidation		53% Oxidation		87% Oxidation	
Day	Average Based on COD	Range	Average Based on COD	Range	Average Based on TOC	Range	Average Based on TOC	Range
2	41	41	91	79 – 100	64	60 – 68	77	70 – 85
4	86	80 – 92	100	100	71	69 – 72	97	95 – 99
6	100	100	100	100	95	95	100	100
9	58	58 – 59	100	100	72	69 – 75	100	100
11	45	39 – 50	75	68 – 83	65	62 – 68	70	63 – 76
13	66	47 - 85	92	82 - 100	80	64 – 97	98	79 – 100

It is apparent that oxidation of DCDE by Fenton can produce by-products that are more biodegradable than DCDE. On the other hand, Fenton oxidation may create by-products which are toxic as seen from long-term respiration, toxicity, and SBR experiments. No noticeable acclimation to the by-products of oxidation was observed in the SBR experiments with Fenton.

7.2.3 UV/H_2O_2 -Treated DCDE

The results of Experiment 1 for UV/H_2O_2 treated DCDE are presented in Figures 7-11 to 7-13. Oxidized chemical was introduced to the reactors on day zero. As seen from Figure 7-11, the MLSS ranged from approximately 2,500 to 5,000 mg/L in the reactors receiving oxidized DCDE. The MLSS was relatively constant at about 3,000 mg/L in the controls. The large MLSS fluctuations in the reactors could be a reflection of biomass reaction to receiving a feed that is 75% by volume of oxidized DCDE.

In this concentrated feed experiment, the COD removal in the reactors receiving chemical was substantially lower than the controls (Figure 7-12). The error in COD measurements was low, and no overlap of data was observed after day 5 between the controls and the reactors receiving oxidized chemical. This suggests that the reactors with chemical were, in fact, less efficient than the controls.

Initially, TOC removal in the reactors receiving oxidized chemical was about 7% lower than the controls, as seen in Figure 7-13. There was a small range of error and no overlap for the TOC removal measurements, suggesting a distinct difference in performance between the reactors with chemical and the controls. The TOC removal increased from 87% to 93% from day 8 to day 15. Over the same period, the COD removal increased from about 85% to 90%. In other words, both COD and TOC removal efficiencies increased with time. It should be noted that the by-products of oxidation from UV/H_2O_2 were found non-toxic, based on the toxicity tests. Also, the long-term experiments showed that by-products obtained by UV/H_2O_2 oxidation of DCDE were not toxic at 100% oxidation. Therefore, the observation of improved biodegradability over time can be explained by the growth of microorganisms that can biodegrade the by-products more readily.

Table 7-5 shows the percent biodegradation calculated for this experiment. The percent biodegradation was calculated starting at day 9 because this is the time when the reactors appeared to attain stability. The percent biodegradation was initially low, but then increased up to an average of 82% based on TOC data. The mid-term experiments, on the other hand, showed that only about 25% of the by-products obtained from DCDE oxidized by UV/H_2O_2 at 100% biodegraded within 6 hours.

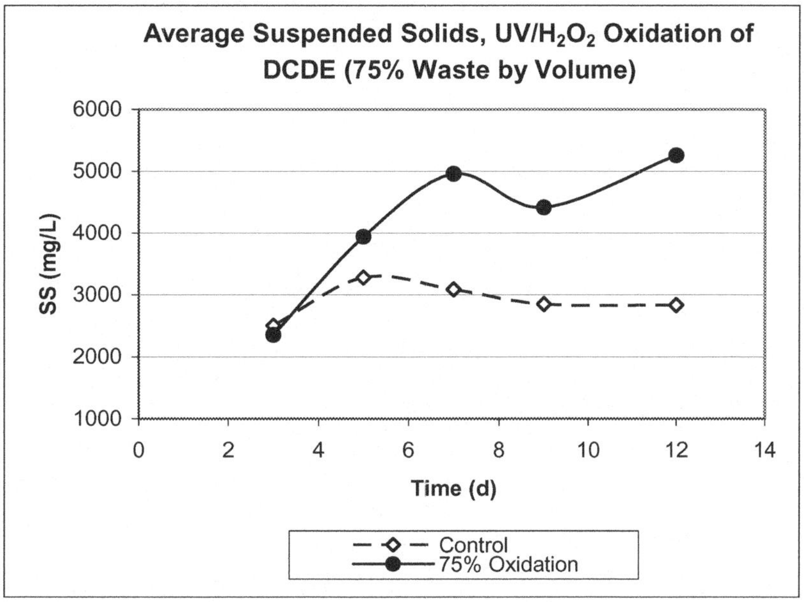

Figure 7-11. Average Suspended Solids vs. Time. (Experiment 1)

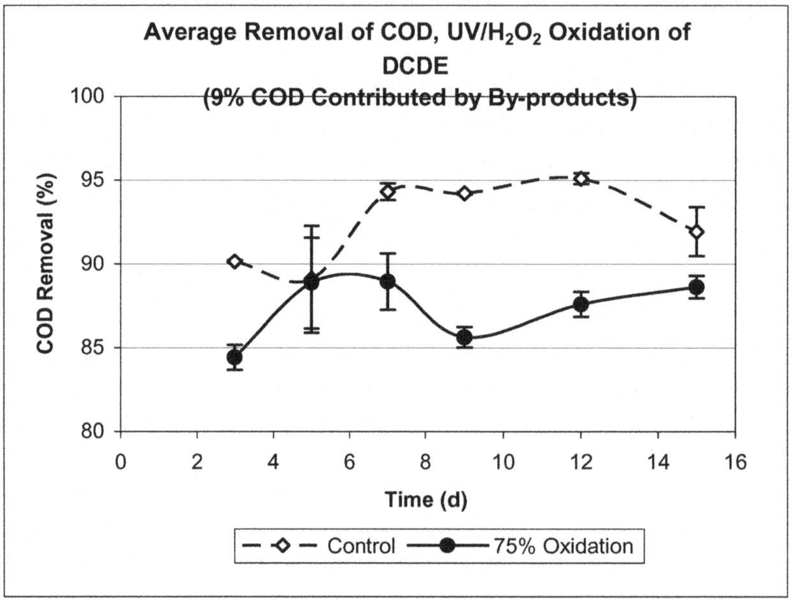

Figure 7-12. Average COD Removal Over Time. (Experiment 1)

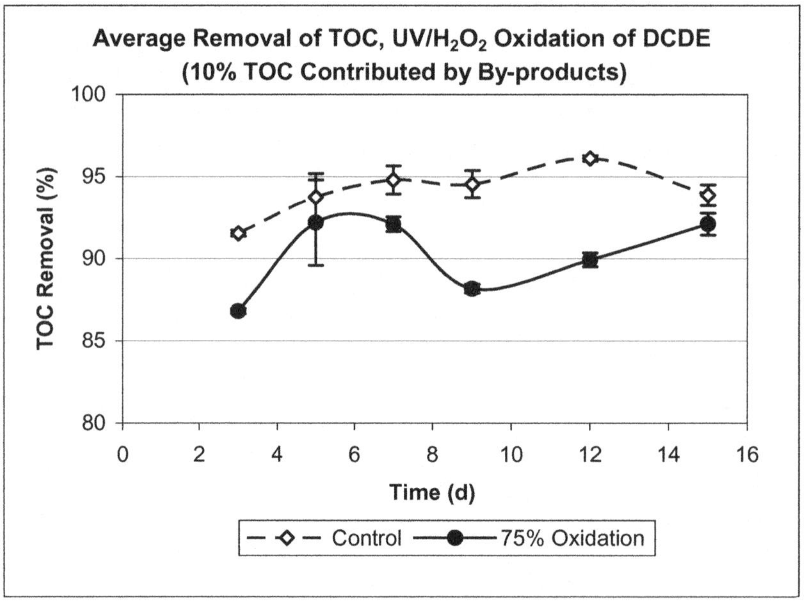

Figure 7-13. Average TOC Removal Over Time. (Experiment 1)

Table 7-5. Estimation of Percent Biodegradation of UV/H_2O_2 Oxidation By-Products of DCDE. (Experiment 1)

By-Product Biodegradation (%)				
Day	Average Based on COD	Range	Average Based on TOC	Range
9	0	0	24	22 – 25
12	18	15 – 21	34	31 – 37
15	61	57 – 66	82	76 – 88

The results with this dilute feed solution (Experiment 2) are presented in Figures 7-14 to 7-16.

The MLSS concentration was maintained between 1,000-2,000 mg/L (Figure 7-14). The biomass concentration was lower in the dilute experiments as compared to the concentrated experiments. This was expected due to a lower COD in the feed solution. Furthermore, the fluctuation of biomass concentration was better controlled in these experiments, probably because only 40% of the feed volume (as opposed to 75%) was contributed by oxidized DCDE.

The reactors containing by-products of DCDE oxidized at 100% showed about 10% to 20% less COD removal, as compared to the controls. Relatively large variations in the COD measurements were observed in these experiments. Again, the oxidation by-products in the samples might have caused interferences with COD analysis. In fact, this effect is more pronounced in the concentrated experiment (Figure 7-15).

The TOC removal seen in the reactors receiving oxidized DCDE was about 7% lower than the TOC removal observed in the control reactors (Figure 7-16). This difference in TOC removal efficiency is, again, probably due to products that are less biodegradable, since these by-products were found to be nontoxic in the earlier experiments.

Table 7-6 shows the percent biodegradation calculated for this experiment. Since the COD data showed a large variation, biodegradation was based on TOC reduction values. As seen in Table 3-3, the average biodegradation was 77%. This value is not very different than the average value of 82% estimated for the concentrated experiment (Experiment 1). However, the major observed difference between the dilute and concentrated experiments was the relatively constant TOC removal throughout the dilute experiment. In contrast, the TOC removal was low at the beginning of the concentrated experiment, but continuously increased with time. The initial low efficiencies in COD and TOC reduction for the concentrated experiment could have been due to the presence of unquenched H_2O_2 in the oxidized samples. For the concentrated experiment, 75% of the feed by volume was contributed by the oxidized solution. In this high volume of oxidized chemical, even small amounts of unquenched H_2O_2 might show toxic effects on activated sludge. On the other hand, the microorganisms in Experiment 1 could have been shocked by receiving a high percentage of oxidized DCDE (75% by volume in Experiment 1 as opposed to 40% by volume in Experiment 2). However, it is apparent that the microorganisms showed rapid acclimation to these conditions.

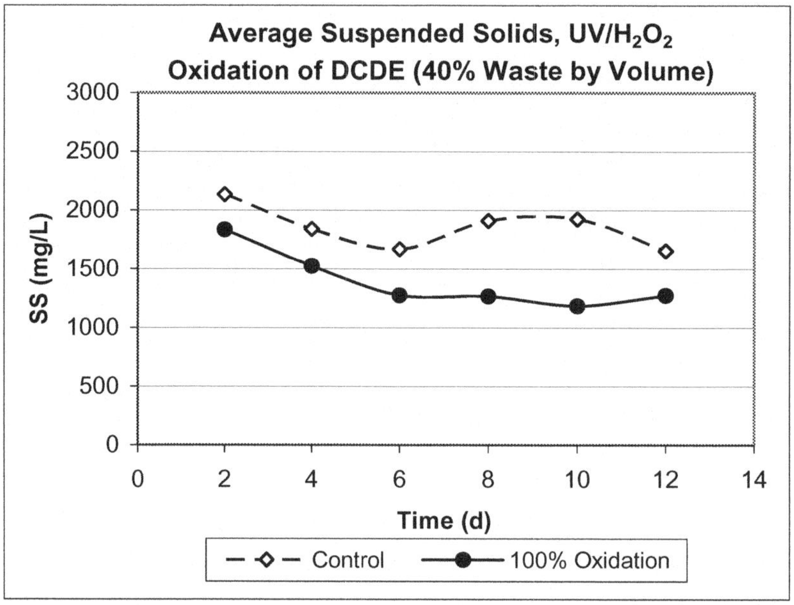

Figure 7-14. Average Suspended Solids vs. Time. (Experiment 2)

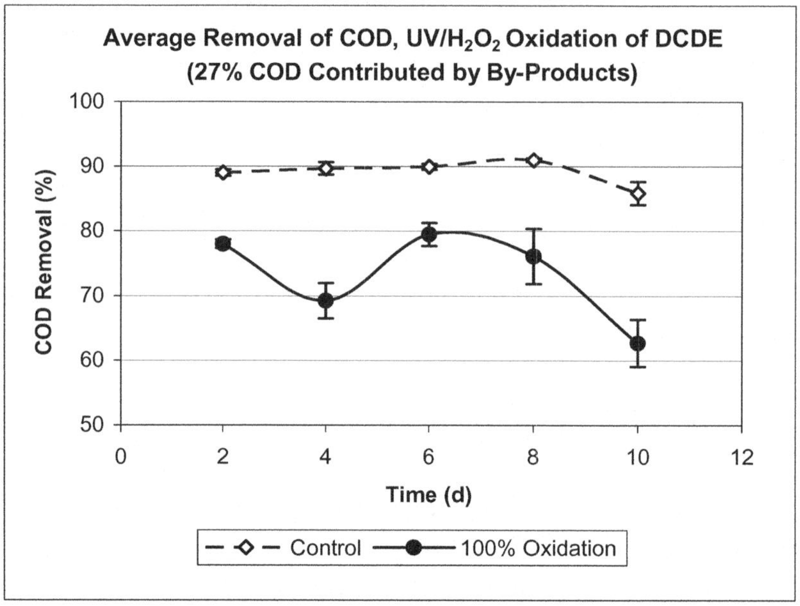

Figure 7-15. COD Removal Over Time. (Experiment 2)

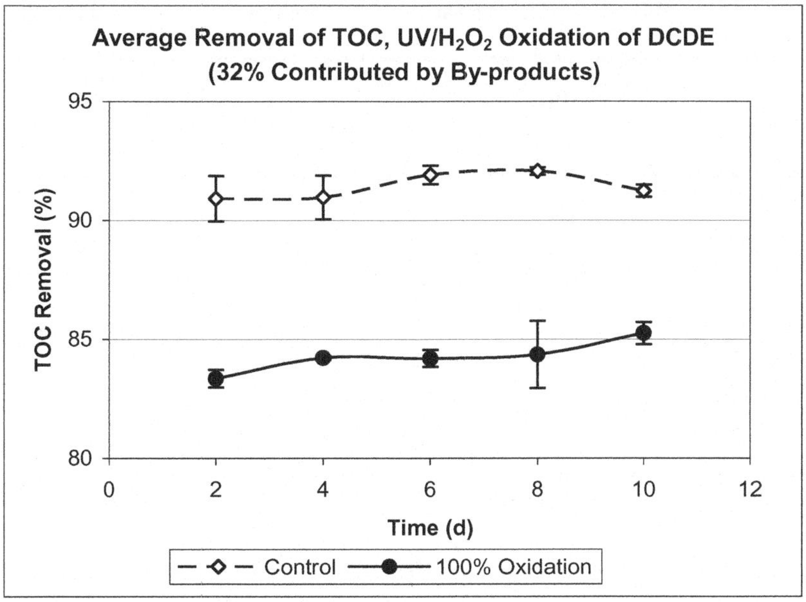

Figure 7-16. TOC Removal Over Time. (Experiment 2)

Table 7-6. Estimation of Percent Biodegradation of UV/H_2O_2 Oxidation By-Products of DCDE. (Experiment 2)

By-Product Biodegradation (%)				
Day	Average Based on COD	Range	Average Based on TOC	Range
2	64	64 – 66	75	75 – 76
4	–	–	78	78
6	65	59 – 70	75	74 – 76
8	33	19 – 47	76	73 – 80
10	–	–	82	80 – 83

The results on UV/H_2O_2 oxidized DCDE showed that about 80% of the oxidation by-products were biodegradable in the SBRs. So, it is obvious that the 24-hour tests fail to reflect the acclimation that can take place in continuously operating systems.

7.2.4 Comparison of Treatments for DCDE

Table 7-7 presents an overall summary for percent biodegradation of DCDE by-products resulting from three oxidation techniques.

Table 7-7. Comparison of Biodegradation Based on TOC Reduction for DCDE Utilizing SBRs.

Test	Oxidation Technique	Oxidation Percentage (%)	By-product Biodegradation Range (%)
Concentrated Feed	Ozone	75	55 – 89
Dilute Feed	Ozone	100	56 – 82
Concentrated Feed	Fenton	53 87	64 – 97 70 – 100
Concentrated Feed	UV/H2O2	75	22 – 88
Dilute Feed	UV/H2O2	100	75 – 83

Overall, all three oxidation techniques produced by-products that were highly biodegradable. After the acclimation periods, about 80–90% of the by-products of ozone and UV/H_2O_2 pretreatment of DCDE was biodegraded. By-products of the Fenton treatment seem to biodegrade more, probably because of excessively high biomass, which accumulated in the reactors due to iron flocculation. Because of steady increase in percent of biodegradation over time, only the range and no average values were presented in Table 7-7.

7.2.5 SBR Tests with Non-oxidized DCDE

Additional experiments were performed with non-oxidized DCDE in order to determine if DCDE had an inhibitive effect on the activated sludge and if it will biodegrade. The DCDE concentration in the feed solution was adjusted to 50 mg/L. Inhibition tests performed earlier with DCDE showed minimum inhibition (about 5%) at a DCDE concentration of 50 mg/L.

In Figures 7-17 and 7-18, the "DCDE" represents the reactor that received 50 mg/L of DCDE in the feed, and the "Control" represents the reactor that received no DCDE or oxidized by-products. The removal efficiencies for COD and TOC removal ranged between 92–95% for the controls, and the DCDE-fed reactors showed removals between 90–94%. Approximately 2% of the COD and 2.5% of the TOC was contributed by DCDE, which was determined to be nonbiodegradable in earlier tests. The DCDE performed similarly to the controls and there was no further inhibition noticed.

The concentration of DCDE in the influent and effluent was measured to determine the extent of DCDE removal. Figure 7-19 shows about 30–40% reduction in DCDE concentration. This reduction is most likely due to adsorption onto biomass because DCDE was observed to be nonbiodegradable in all other tests. Furthermore, DCDE did not volatilize in experiments conducted without biomass.

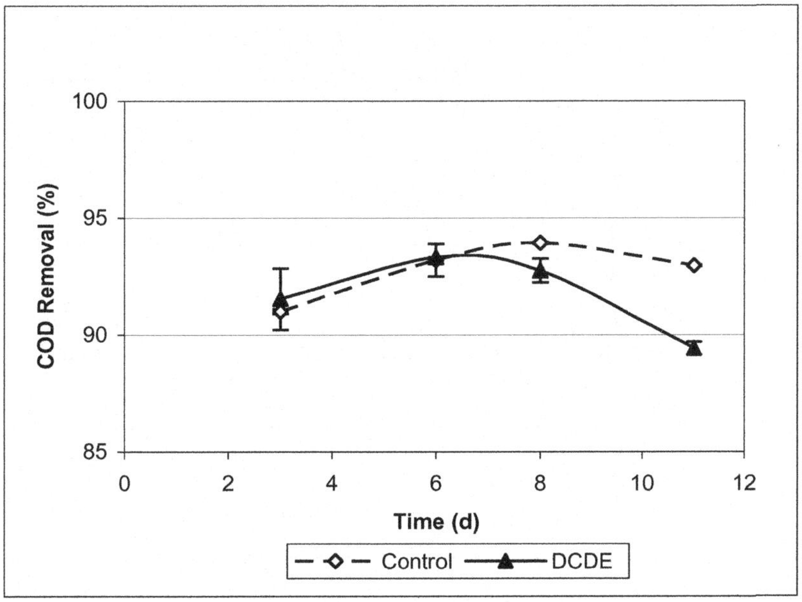

Figure 7-17. COD Removal in DCDE-Fed Reactors.

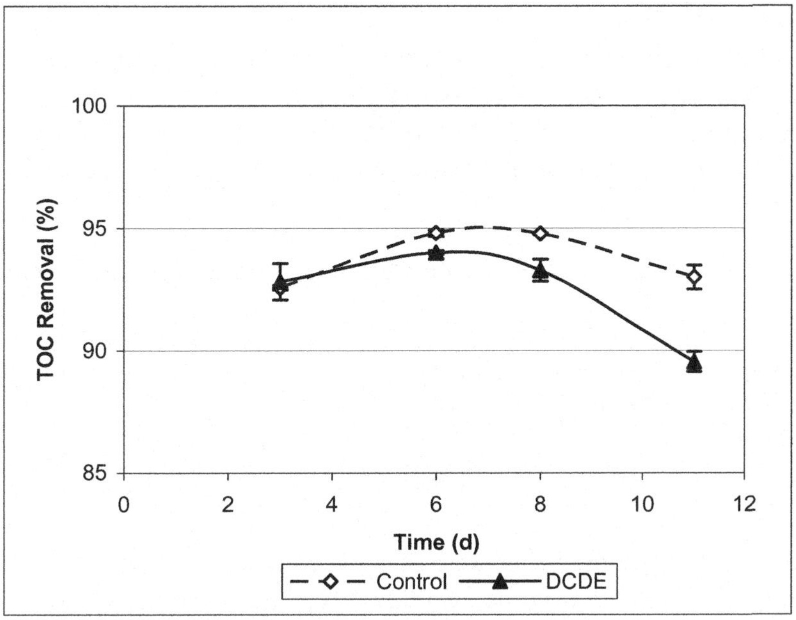

Figure 7-18. TOC Removal in DCDE-Fed Reactors.

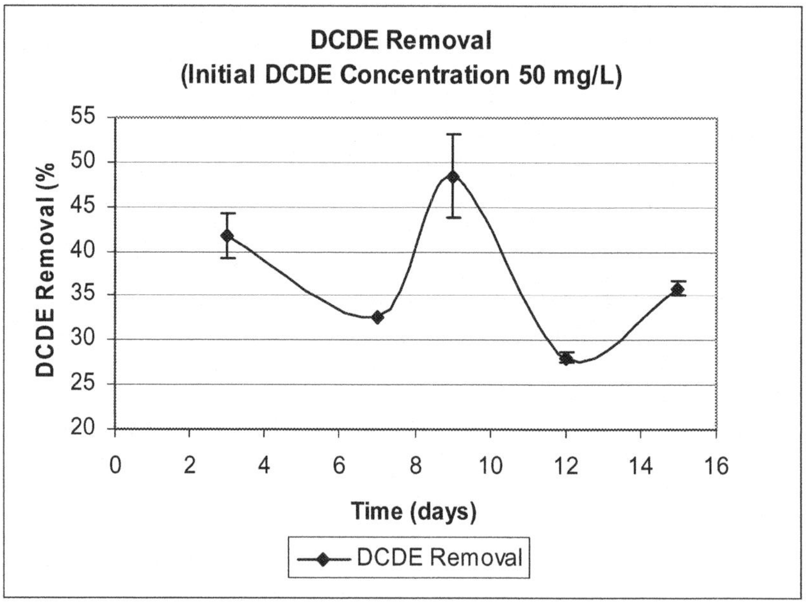

Figure 7-19. DCDE Reduction vs. Time.

7.3 Tests with MTBE

Table 7-8 shows the target contributions of MTBE by-products to COD and TOC for each experiment, as well as other target parameters for each experiment. For MTBE, the only oxidation method tried was UV/H_2O_2 due to significant loss of MTBE by stripping during ozonation and Fenton treatment, as presented earlier (Sections 4.1.2 and 4.2.2).

Table.7-8. Target COD and TOC Contributions of Oxidized MTBE to the Feed Solution.

Experiment No.	Chemical	Oxidative Treatment	Total COD Input	Waste Volume	Oxidation Percentage	TOC Contribution	COD Contribution
10	MTBE	UV/H_2O_2	1500	75%	75%	15%	21%
11	MTBE	UV/H_2O_2	1500	75%	100%	20%	28%
12	MTBE	UV/H_2O_2	300	16%	100%	19%	28%

The results of Experiments 10 and 11 with concentrated feed are presented below in Figures 7-20 to 7-22. The data obtained on day 2 represent the initial conditions of the reactors without chemical addition to the feed solution. Oxidized MTBE solution was added to the feed solution on day 3. The first sample after addition of oxidized MTBE was taken on day 4.

As seen in Figure 7-20, the MLSS concentration ranged from 3500 to 5500 mg/L for this experiment. Solids wasting was performed if the concentration of solids exceeded 4500 mg/L. High fluctuations in the MLSS values were most likely due to difficulty of biomass to adjust to high volumes (75% by volume) of oxidized MTBE solution added.

There was a significant difference in reactor performances between the controls and the reactors receiving 75% and 100% oxidized MTBE. However, the difference between the responses of the two reactors receiving 75% and 100% oxidized MTBE was negligible. This was in spite of significant differences in MLSS levels in these reactors. There was also a decrease in COD and TOC removal efficiencies for all reactors towards the end of the experiment. This decrease is obviously not due to oxidized MTBE.

The reactors receiving oxidized chemical consistently performed less efficiently than the controls. The average difference in efficiencies was 10% for COD and 7% for TOC reduction. This difference is probably due to incomplete biodegradation of the by-products rather than inhibition. Inhibition tests showed that MTBE oxidized by UV/H_2O_2 caused no inhibition.

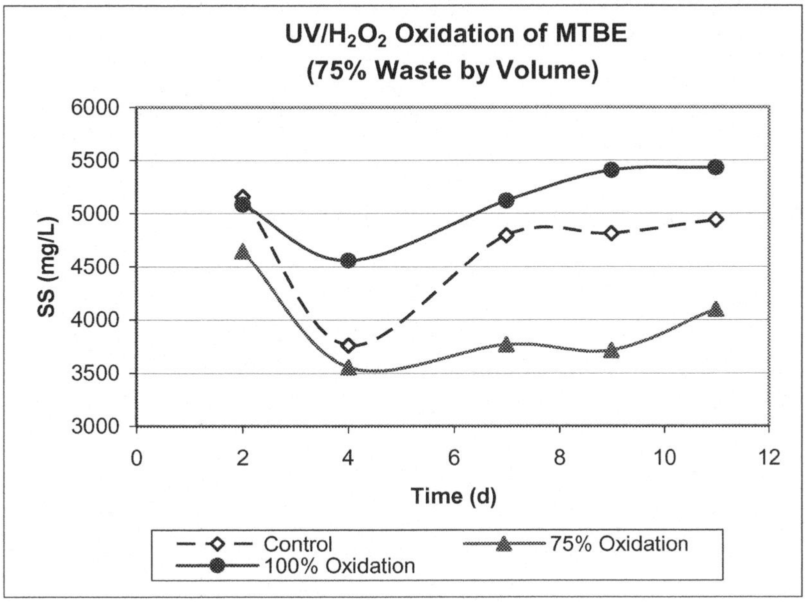

Figure 7-20. Average Suspended Solids vs. Time. (Experiments 10 and 11)

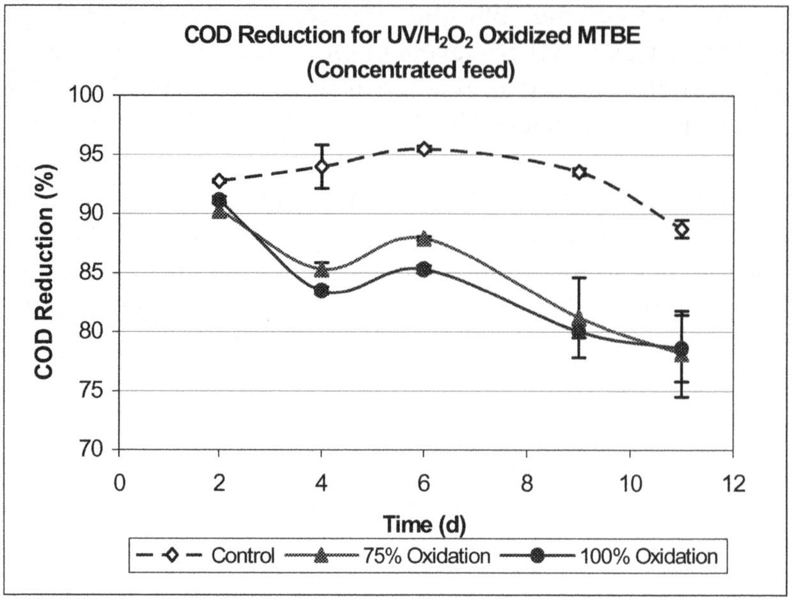

Figure 7-21. Average COD Removal Over Time. (Experiments 10 and 11)

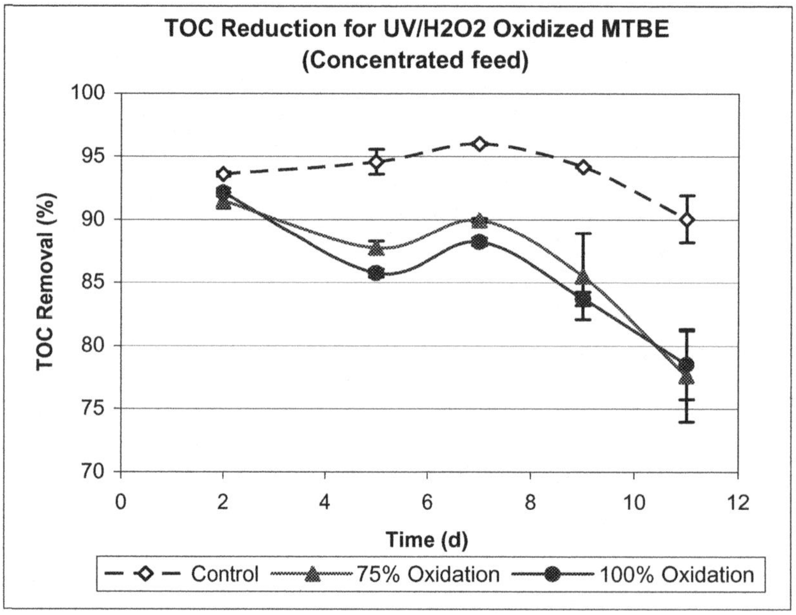

Figure 7-22. Average TOC Removal Over Time. (Experiments 10 and 11)

Table 7-9 shows the calculated biodegradation of MTBE by-products for Experiments 10 and 11. Based on COD data about 60% of the by-products of MTBE oxidation by UV/H_2O_2 was biodegradable. The TOC data indicated about 50% biodegradation of the by-products. As opposed to the results with oxidized DCDE, no acclimation over time was noticed for these experiments.

Table 7-9. Estimation of Percent Biodegradation of UV/H_2O_2 Oxidation By-Products of MTBE. (Experiments 10 and 11)

	By-Product Biodegradation (%)							
	75% Oxidation		100% Oxidation		75% Oxidation		100% Oxidation	
Day	Average Based on COD	Range	Average Based on COD	Range	Average Based on TOC	Range	Average Based on TOC	Range
4	61	59 – 63	62	62 – 63	52	51 – 55	51	48 – 54
7	57	56 – 57	55	54 – 56	49	48 – 50	47	47 – 48
9	59	59	54	53 – 55	50	50	43	42 – 43
11	54	58 – 59	68	60 – 75	53	53	48	48

Figures 7-23 to 7-25 show the results of Experiment 12 with the dilute feed solution of 300 mg/L of COD. The MTBE by-products contribute approximately the same amount of COD and TOC as the previous two experiments because of reduction in volume contribution to 16%. The target oxidation of MTBE was 100%, but sample analysis after oxidation indicated that MTBE was oxidized by only 80%. The oxidized MTBE was added to the feed solution on day zero.

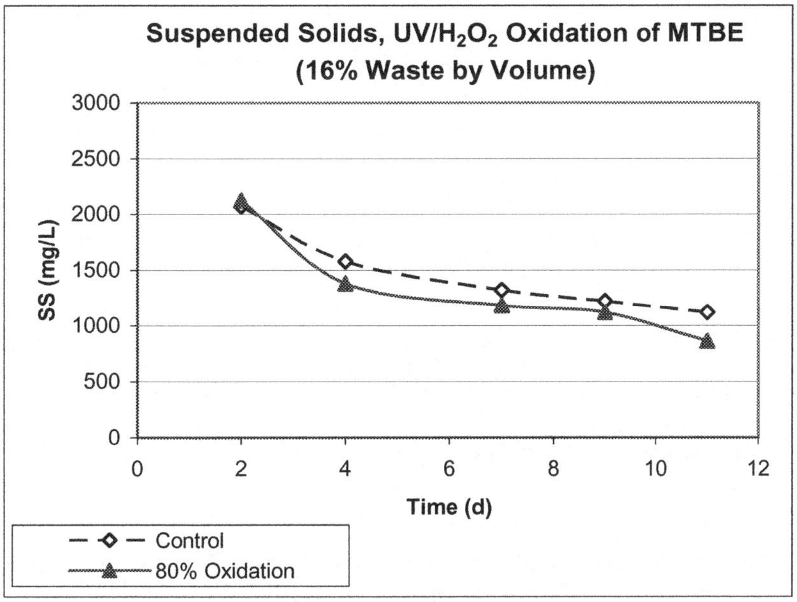

Figure 7-23. Average Suspended Solids vs. Time. (Experiment 12)

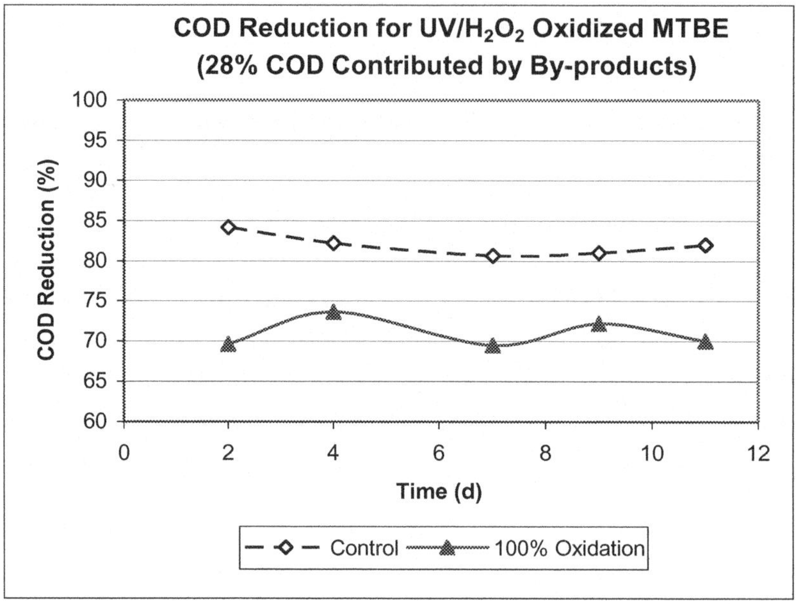

Figure 7-24. Average COD Removal. (Experiment 12)

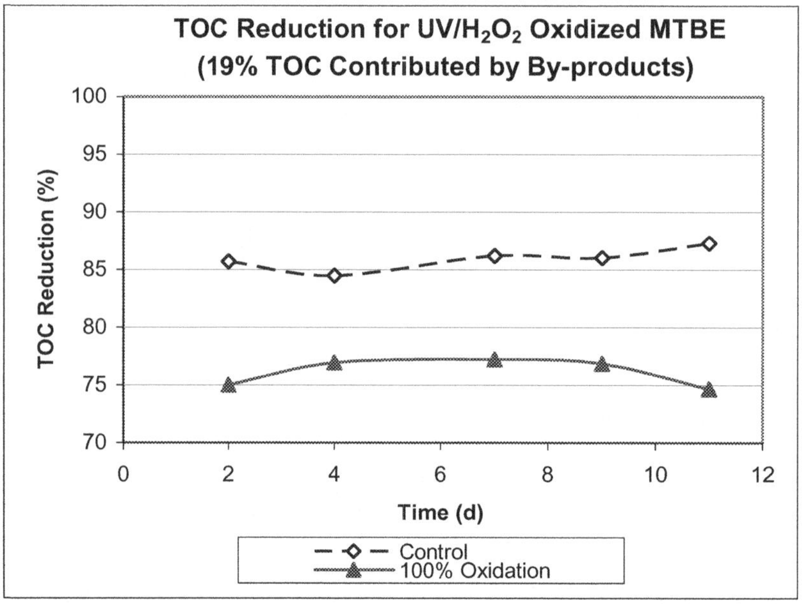

Figure 7-25. Average TOC Removal. (Experiment 12)

The MLSS ranged between 1,000–2,000 mg/L. There was not a significant difference between the solids concentration of the reactors with oxidized chemical as compared to the control, again due to low levels of volume contribution (165) by the oxidized MTBE.

The COD removal efficiency in the control reactor was about 81–84%, although the efficiency of the reactors receiving oxidized MTBE varied from 68–74%, representing about a

10% reduction in COD for the reactors receiving MTBE by-products. Similarly, about 10% reduction in TOC performance was observed compared to control. The standard error bars were not included in Figures 8-24 and 8-25, because the standard deviation was very low (±1%).

In comparing the control reactors in Experiments 10, 11, and 12, i.e., the reactors receiving concentrated versus dilute feed without any oxidized MTBE, the following observations can be made: The control reactors receiving the dilute feed produced about 85% removal of the COD and 85% removal of TOC. This was approximately 10% lower removal of COD and TOC as compared to the performance of the control reactors receiving concentrated feed, which averaged at about 95% for COD and TOC removal. This decrease is most likely due to reduced amount of food, which resulted in a MLSS concentration of 1000-2000 mg/L in reactors receiving dilute feed compared to 4,000-5,000 mg/L of MLSS in the reactors receiving the concentrated feed.

Table 7-10 presents the percent biodegradation of the by-products estimated for Experiment 12. The percent biodegradation results are similar to the results obtained from the previous concentrated feed experiment (see Table 7-11 for comparison). The average biodegradation was estimated to be 62% as measured by COD and 48% as measured by TOC. It should be noted that the mid-term experiments resulted in average npCOD and npTOC reductions of 57%. Therefore, it can be concluded that there was no significant difference between percent biodegradation results obtained by the mid-term tests and the SBR experiments.

Table 7-10. Estimation of Percent Biodegradation of UV/H_2O_2 Oxidation By-Products of MTBE. (Experiment 12)

By-Product Biodegradation (%)				
Day	**Average Based on COD**	**Range**	**Average Based on TOC**	**Range**
2	54	52 – 57	53	53
4	68	66 – 71	53	52 – 53
7	69	68 – 70	47	46 – 47
9	67	64 – 70	51	50 – 53
11	51	50 – 53	38	36 – 40

Table 7-11. Comparison of Biodegradation Based on TOC Reduction for MTBE Utilizing SBRs.

Test	Oxidation Technique	Target Oxidation Percentage (%)	Actual Oxidation Percentage (%)	By-Product Biodegradation Based on COD (%)	By-Product Biodegradation Based on TOC (%)
SBR Concentrated Feed	UV/H_2O_2	75 100	75 100	58 ± 5 60 ± 10	51 ± 3 48 ± 6
SBR Dilute Feed	UV/H_2O_2	100	80	62 ± 10	48 ± 9

CHAPTER 8.0

EXPERIMENTS WITH GROUNDWATER AND WASTEWATER

8.1 MTBE-Containing Groundwater

8.1.1 Characteristics and Chemical Oxidation

Groundwater was obtained from BP Carson Refinery in Los Angeles County, California. The site and the groundwater are known to be contaminated by petroleum components. The major parameters measured in groundwater in the project team's laboratories are listed in Table 8-1.

Table 8-1. Groundwater Characteristics.

Parameter	Value
pH	6.7
Turbidity (NTU)	15
Total Alkalinity (mg/L as $CaCO_3$)	560
Methyl tert-butyl ether (MTBE) (mg/L)	610
Benzene (mg/L)	7
Toluene(mg/L)	17
tert-Butyl alcohol (TBA) (mg/L)	11
TOC (mg/L)	490
COD (mg/L)	1900

A sample of groundwater was oxidized by using the UV/H_2O_2 process without filtration to remove the turbidity. The same oxidation setup and the procedure explained for MTBE solutions were used without any change: 0.97 mmol H_2O_2 was added to 3 L solution at 5-minute intervals. Because the main pollutant in the groundwater was MTBE (85% of TOC and 88% of COD), MTBE concentration was monitored during oxidation to check the oxidation level. The change in MTBE concentration with time is shown in Figure 8-1.

For synthetically prepared MTBE solution at 249 mg/L concentration, this procedure resulted in about 91% removal of MTBE within 90 minutes, which needed a total of 17.5 mmole H_2O_2 (See Figure 4-20). In groundwater, the same procedure provided only 80% oxidation in 6 hours and 64 mmole H_2O_2 addition (Figure 8-1). There are several reasons for this discrepancy:

1. The groundwater contained 2.5 times more MTBE than the synthetic solutions;
2. There was about 75 mg/L of TOC and 240 mg/L of COD containing chemicals in the groundwater other than MTBE, e.g., benzene, toluene, and TBA);
3. The groundwater sample was turbid, and therefore, the light intensity is expected to be diminished considerably while applying the UV/H_2O_2 technique to groundwater; and
4. The alkalinity of the synthetic MTBE solution was 250 mg/L as $CaCO_3$, adjusted by using 5 mM bicarbonate ion.

The groundwater had more than twice as much alkalinity, or about 11 mM bicarbonate ion. Bicarbonate and carbonate ions, being well-known hydroxyl radical scavengers, would compete successfully with the targeted organic compounds.

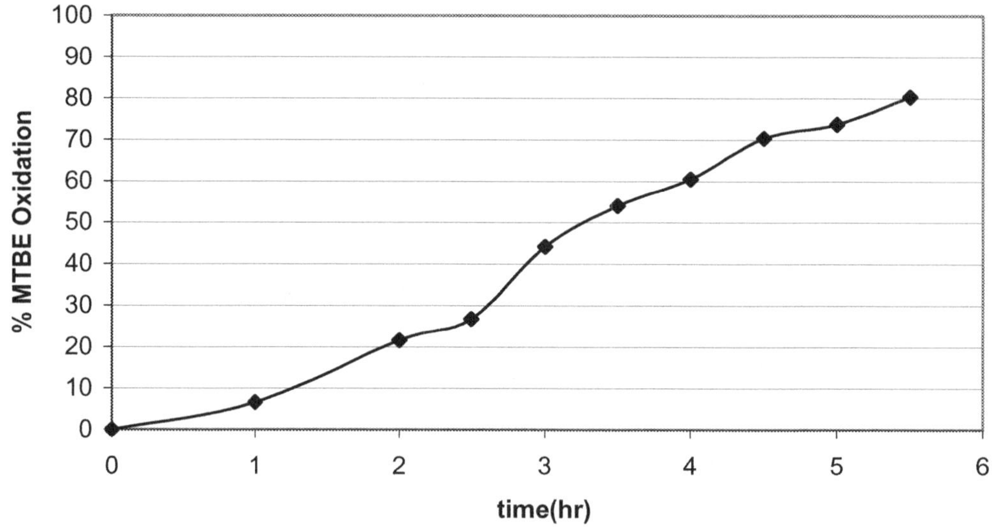

Figure 8-1. Change in MTBE Concentration with Time During Groundwater Oxidation with UV/H_2O_2 Technique.

8.1.2 Long-Term Respirometric Tests

Respirometry studies were performed with groundwater before oxidation and after 70% and 80% oxidation of MTBE by UV/H_2O_2. Both acclimated and non-acclimated activated sludge cultures were used in the tests. The TOC values of the samples were measured before and after they were subjected to the respirometric tests. The results are presented in Table 8-2.

Table 8-2. Change in TOC Values During Respirometer Experiment (189 hours).

	Initial TOC (mg/L)	Final TOC (mg/L)	TOC Change (mg/L)	% TOC Decrease
Groundwater	220	35	185	84.1
70% Ox. GW-Non-A	191	68	123	64.4
80% Ox. GW-Acc	190	73	117	61.6

The TOC data indicate that all of the samples tested went through a significant amount of biodegradation. Some of the TOC reduction is probably due to the stripping of organic matter during respirometric experiments. It should be noted that the groundwater sample tested here had only 220 mg/L of TOC, which is considerably lower than the TOC shown in Table 8-1, because the original and the oxidized samples were diluted by half before respirometric studies. Furthermore, the groundwater contained high concentrations of MTBE, which is quite volatile. Chemical oxidation of the MTBE by 70% and 80% by UV/H_2O_2 reduced the TOC by about 14% to 190 mg/L, partially by stripping and partially by oxidation of MTBE. However the TOC reduction during the respirometric study of the oxidized groundwater is about 62-64%. These samples have much less MTBE to volatilize during the respirometric study. Oxidation by-products of MTBE, e.g., TBA, are not volatile. So, it can be assumed that this reduction in TOC was due to chemical oxidation primarily.

Net oxygen consumption and net carbon dioxide production data are presented in Figures 8-2 and 8-3.

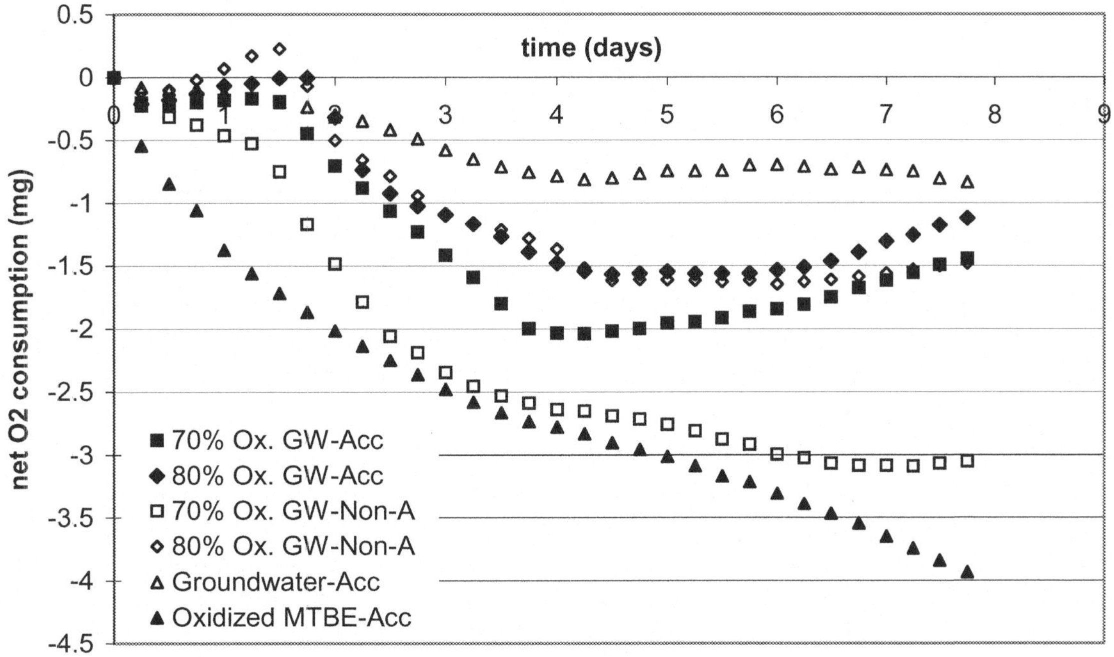

Figure 8-2. Net Oxygen Decline for Respirometric Experiments with Groundwater.

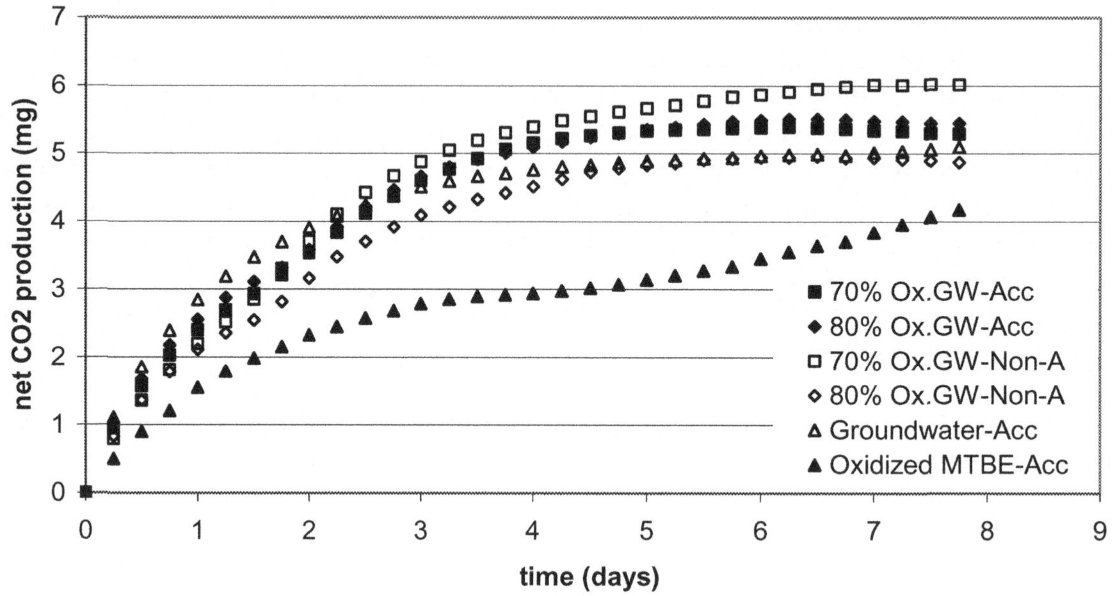

Figure 8-3. Net CO_2 Production Results for Respirometric Experiments with Groundwater.

The non-oxidized groundwater consumed the least amount of oxygen, as expected. According to Figure 8-2, after chemical oxidation, microorganisms were able to more readily biodegrade the oxidation by-products. This is consistent with our earlier observations on MTBE samples (Section 5.2.3). The results of CO_2 production did not distinguish the different treatments well, as shown in Figure 8-3. However, the sample oxidized by 70% and then subjected to non-acclimated culture seemed to create higher respirometric rate than the others.

8.1.3 SBR Tests on Groundwater Samples

The SBR test was conducted on UV/H_2O_2-treated BP groundwater. The experiment involved one control reactor that received no groundwater, one reactor that received non-oxidized groundwater, and two reactors that received groundwater treated by UV/H_2O_2 to the level of 80% MTBE removal. The total COD of each feed was 300 mg/L. The two controls received only the Syntho solution with a COD of 300 mg/L. In the reactors that received oxidized or non-oxidized groundwater, 30% of the COD and TOC was contributed by the organic matter in groundwater to the feed solution.

As seen from Figure 8-4, the suspended solids ranged from 1,000–2,000 mg/L. There was not much difference between the biomass concentrations of the control and the other reactors. Figure 8-5 shows that COD removal in the reactor that received non-oxidized groundwater was 5–10% less than the COD removal in the control reactor. The two reactors that received oxidized groundwater produced even less COD removal.

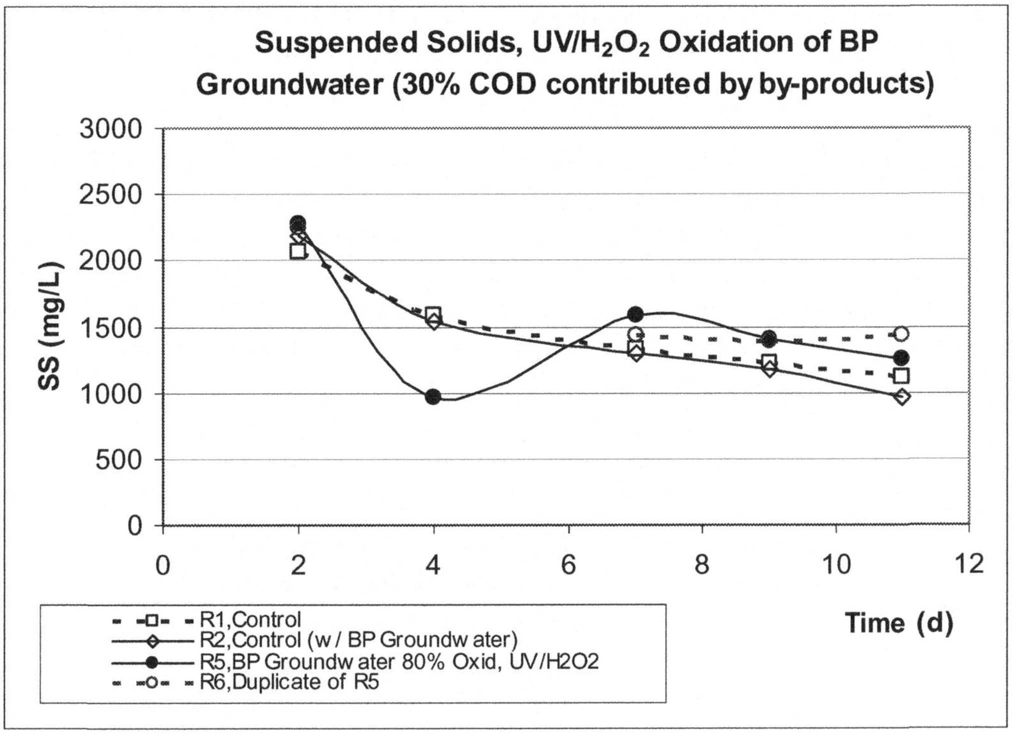

Figure 8-4. Suspended Solids vs. Time.

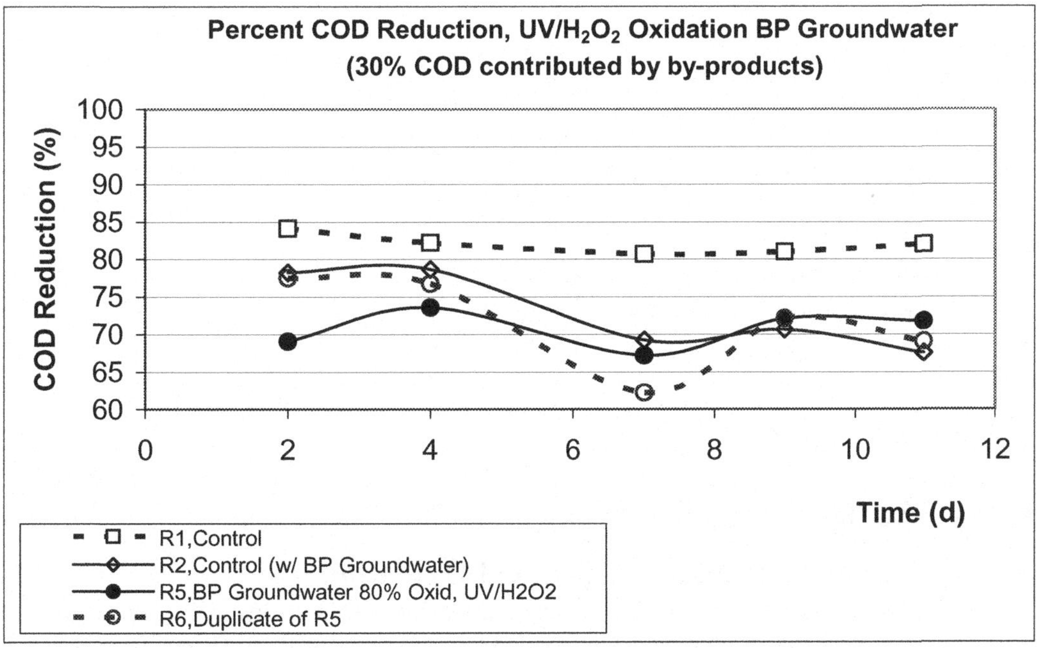

Figure 8-5. COD Removal Over Time.

The TOC reduction in SBRs is shown in Figure 8-6. The control reactor and the reactor that received non-oxidized groundwater performed similarly. It should be noted that groundwater was added to the SBR at a dilution of 35 ml per 280 ml of the working volume. The MTBE at a concentration of 610 when diluted at this rate will result in 76 mg/L of MTBE in the SBR. It is highly likely that most of this MTBE was removed by volatilization during the experimentation.

The TOC removal in the reactors that received oxidized groundwater was about 3–4% less than the TOC removal observed for the control reactor. That translates into 85–90% by-product biodegradation estimated based on the procedure described in Section 7.1. It should be noted that these values are larger than the 62–64% observed during the long-term respirometric study, probably because of higher chance of volatilization of remaining MTBE in an SBR than in the respirometer. It should be noted that MTBE was oxidized to the level of 80%, still leaving about 120 mg/L of MTBE in solution after oxidation. At the dilution rate applied, the SBR contained about 15 mg/L of MTBE.

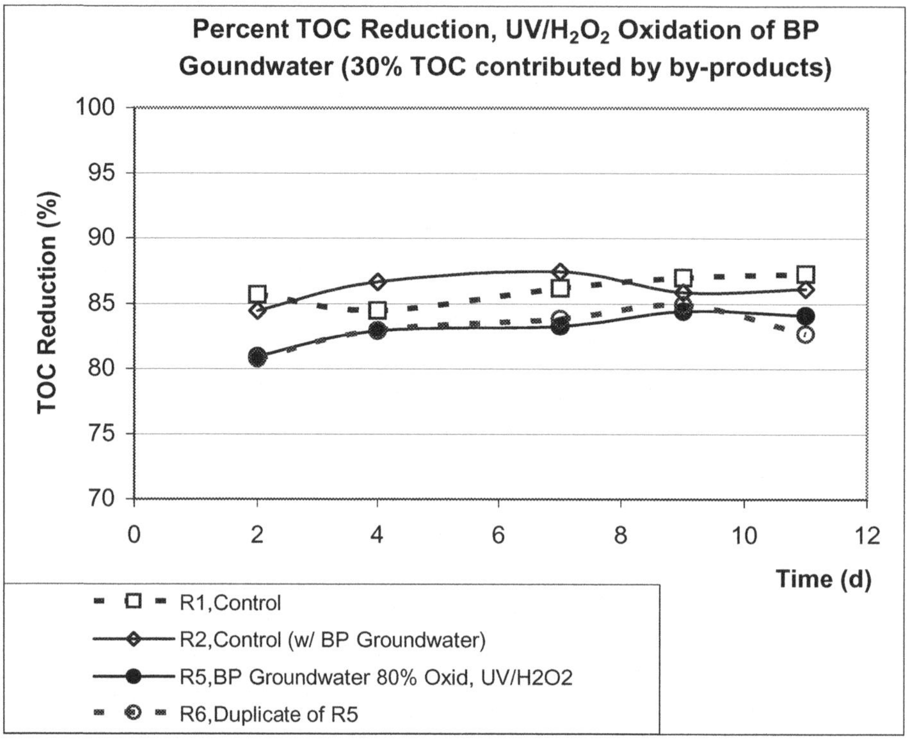

Figure 8-6. TOC Removal Over Time.

8.2 Hollow Fiber Membrane Waste

8.2.1 Waste Characterization and Chemical Oxidation

The second wastewater was obtained from a hollow fiber membrane manufacturer in San Diego County, California. The phase inversion process used in hollow fiber spinning manufacture produces a waste stream that has a high TOC (14,000 mg/L) and high COD (25,000 mg/L). The waste consists of dimethyl formamide (DMF), n-methylpyrrolidone (NMP), glycerin (Gly), PVP, or an ethoxylate block copolymer Tetronic 1307 or Tetronic 908. Because of the variable nature of the process waste stream, the concentrations in Table 8-3 were used to simulate an average waste (all samples were made up in deionized (DI) water).

Table 8-3. Makeup of Synthetic Waste.

Compounds	Concentration (mg/L)
Dimethyl formamide (DMF)	22,000
n-Methylpyrrolidone (NMP)	1,900
Glycerin (Gly)	300
Polyvinylpyrrolidone (PVP)	190
Tetronic 1307 or 908	500 – 2,000

Currently this waste stream is treated onsite by a membrane bioreactor (MBR). The fiber spinning waste at a flow of 6,000 gal/day constitutes about 25% of the flow to the MBR, the balance comes from the casting process line. The hollow fiber spinning waste causes two operational problems in the MBR: excessive foaming from the surfactants and PVP, and membrane fouling by PVP.

Tetronic and PVP are not biodegradable and are very soluble in water, so biotreatment and coagulation would not be effective. Previous work (Adams and Kuzhikannil, 2000; Horikoshi et al., 2001; Kitis et al., 1999) indicated that both of these problems could be solved using AOP. Because the application of ozone would cause foaming problems and the formation of solid precipitates from Fenton's reagent would interfere with PVP removal, only UV/H_2O_2 was investigated as an AOP.

The oxidation of PVP results in polymer chain scission. As oxidation progresses, the length of the polymer becomes shorter and shorter, decreasing its affinity for the membrane and increasing its biodegradability. The surfactant Tetronic is oxidized in a similar fashion; the hydrophobic chains are shortened, and the addition of OH groups changes the polarity. Thus, foam stability, which depends on surface tension and the formation of bilayers, is changed.

PVP could not be analyzed by GC because of its high molecular weight (~900 kDa). High background concentrations of DMF-NMP made it impossible to monitor PVP concentrations by TOC and COD. The method of Levy and Fergus (1953) was modified to measure PVP by forming a colored complex with iodine. Samples of PVP were diluted and potassium triiodide (0.006 N) was added. The resulting colored complex was measured at 500 nm spectrophotometrically. The method was modified so that dilutions were made in DI water rather than citric acid. This made the color solutions stable in the presence of H_2O_2. One problem observed was that the absorbance changes with the PVP chain length, so that shorter chains give a lower concentration (10%–17%). PVP standards were purchased from TCI America, (Portland OR), PVP K15 MW=1000, PVP K90 MW=360000. Experimental data was calibrated using the PVP supplied by the membrane manufacturer (Hydra) and had an estimated MW=900000. This method also gives no information on chain length, which may be an important parameter in determining membrane fouling and biodegradability. The calibration curve is presented below.

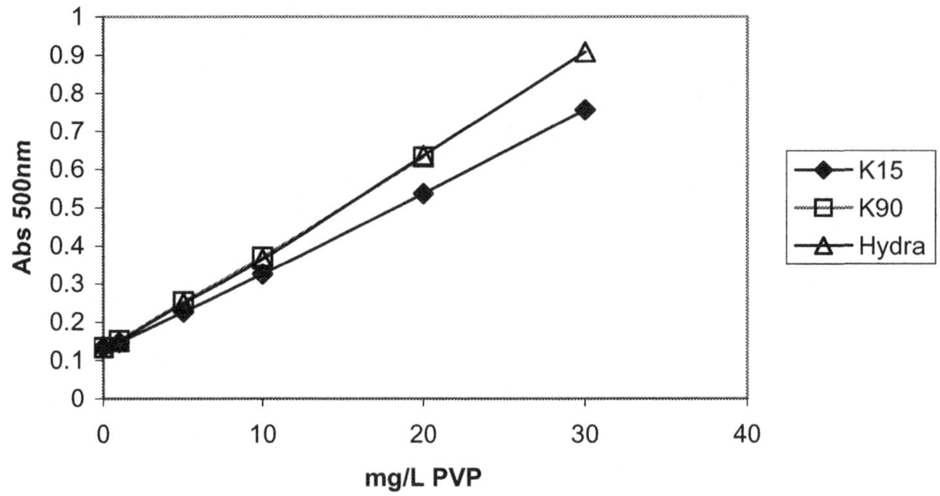

Figure 8-7. PVP-I2 Calibration with PVP of Different Chain Lengths.

Table 8-4 lists the UV/H_2O_2 oxidation experiments performed. Reactions were performed in a 3 L quartz batch reactor in a Rayonet photoreactor equipped with either 8 or 16 low-pressure mercury vapor lamps. These lamps emit UV at a single wavelength, 254 nm. Hydrogen peroxide absorbs UV at this wavelength, but none of the waste components do. The UV intensity was measured by chemical actinometery (iodide-iodate method, Rahn et al., 1999). The reaction matrix was either DI water or DMF-NMP-Gly matrix (Waste) at the concentrations listed in Table 8-3.

Table 8-4. UV/H_2O_2 Oxidation of PVP.

Exp #	UV w/s	[H_2O_2], M	Matrix	PVP % Removal	Biodeg.	Reaction Time (hours)
1	2.53	0.04	DI	100	Enhanced	0.5
2	2.53	0.04	Waste	35	No Effect	1
3	4.86	0.08	Waste	45	Enhanced	3
4	4.86	0.04 x 3	Waste	100	Enhanced	3

Experiment 1 was oxidation of 190 mg/L PVP in DI water. The PVP was completely removed within 30 minutes (Figure 8-8). In the presence of the other waste components (Experiment 2) however, oxidation at the same UV intensity and H_2O_2 concentration only removed 35% of the PVP before the H_2O_2 was depleted.

Table 8-5. Reaction Rates with Hydroxyl Radicals.

Compound	Reaction Rate $M^{-1}s^{-1}$	Initial Concentration, M
PVP as Vinylpyrrolidone	7.5×10^9	0.00171
Dimethyl Formamide	1.7×10^9	0.301
n-Methylpyrrolidone	2.6×10^9	0.032
Glycerol	2.0×10^9	0.00326
Products	1.0×10^9	0

Comparing the relative rates of reaction of PVP and DMF from Table 8-5, it can be seen that only about 2.5% of the available •OH radicals will react with PVP.

$$d[PVP]/dt = k_{PVP} \times [PVP] \times [\bullet OH]$$

$$d[DMF]/dt = k_{DMF} \times [DMF] \times [\bullet OH$$

$$\frac{k_{PVP}[PVP]}{k_{DMF}[DMF]} = 0.025$$

Thus it is expected to take about 40 times longer to oxidize PVP in the DMF-NMP-Gly matrix as it does to treat PVP alone (Figure 8-8).

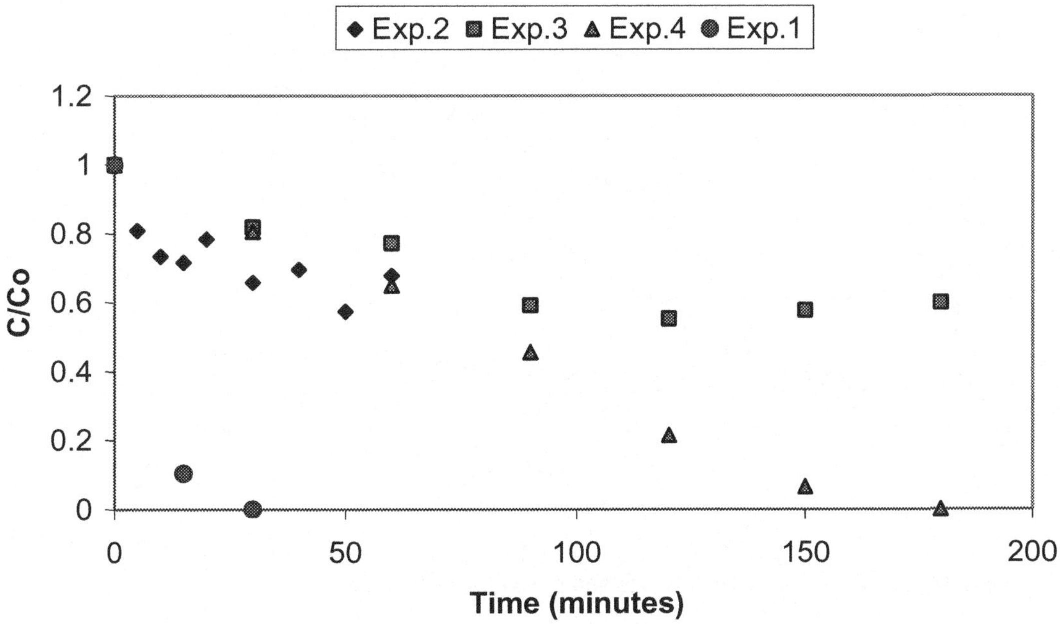

Figure 8-8. PVP Removal by UV/H_2O_2.

In Experiment 3, UV intensity and H_2O_2 concentration were doubled, but the PVP removal was about the same. Analysis of H_2O_2 concentrations in Experiments 2 and 3 showed that when concentrations fell below 20 mM H_2O_2, the removal of PVP stalled. To avoid this, in Experiment 4, serial additions of 40 mM hydrogen peroxide were added at 1-hour intervals, and complete removal was achieved.

8.2.2 Long-Term Respirometric Tests

Long-term respirometric measurements were run using an Oxymax Respirometer (Columbus Instruments) used for other experiments. A sample (25 ml) was seeded with a mixed culture of microorganisms (25 ml sewage), and the consumption of oxygen and the production of carbon dioxide were measured for several days. Initial measurements of the biodegradability of non-oxidized solutions of PVP, DMF, a mixture of DMF-NMP-Gly, and the simulated waste of PVP-DMF-NMP-Gly are presented in Figure 8-9 in terms of carbon dioxide formation. PVP was not biodegradable, DMF degraded slowly, and the mixtures degraded faster (probably due to the presence of glycerin). The Endogenous line is the CO_2 produced by the seed organisms alone. The PVP line falls on top of the endogenous line.

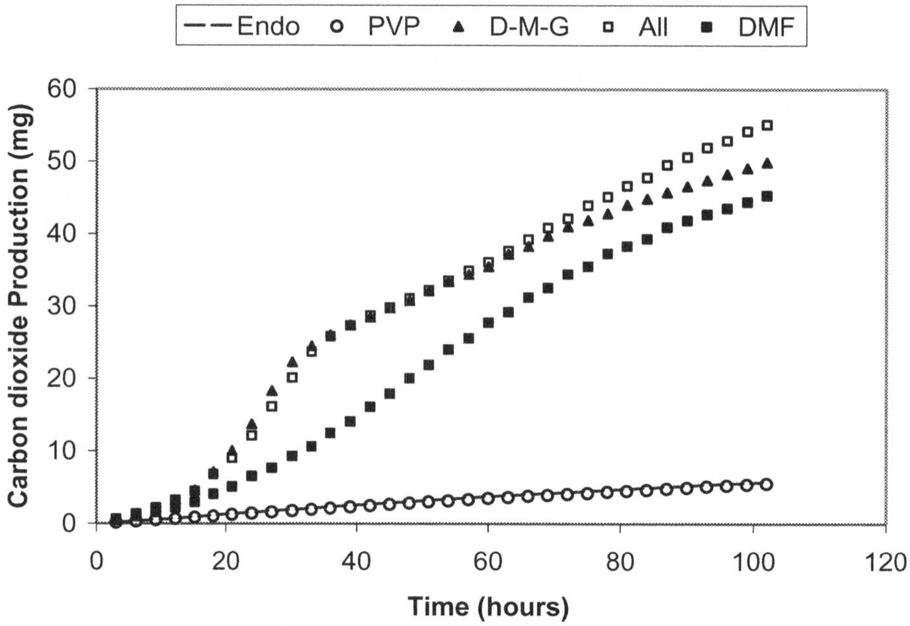

Figure 8-9. Biodegradability of Untreated Waste and Individual Compounds.

Figure 8-10 shows that oxidized PVP is more biodegradable than non-oxidized PVP. In this experiment, the PVP in DI water was preoxidized to remove PVP by 100% (Experiment 1).

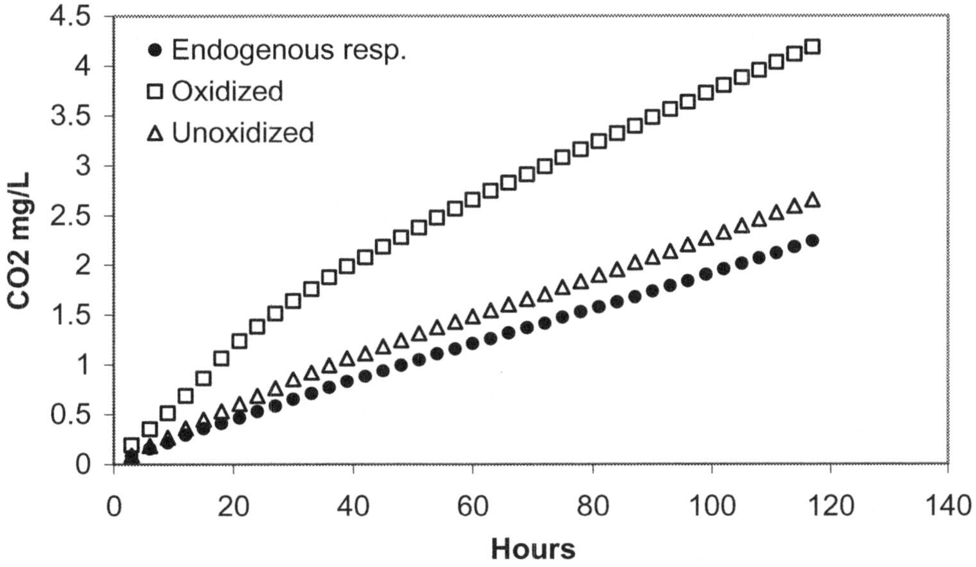

Figure 8-10. Long-Term Respirometric Test Results for PVP Solutions.

Figure 8-11 compares the CO_2 production from non-oxidized waste and oxidized waste (Experiment 3, and Experiment. 4). Both Experiment 3 (partial removal of PVP) and Experiment 4 (100% removal) showed increased biodegradation compared to the non-oxidized waste. As shown in Figures 8.10 and 8.11, the 100% oxidized PVP solution generated about 4,000 μg CO_2, whereas the simulated waste with 100% PVP oxidation produced about 40,000 μg CO_2. So the CO_2 contribution of oxidized PVP is overshadowed by the degradation of the other components. Additional experiments are needed to determine if partial oxidation will increase biodegradation of PVP specifically.

It should be noted that after 4 days the content of Experiment 3 began to degrade at a faster rate. This is indicative of acclimation, where a population of microorganisms capable of utilizing a substrate (possibly DMF) grows large enough to affect the overall rate. Identification of these organisms was not attempted. All eight (each sample was run in duplicates) of these samples were seeded from the same sewage sample. But obviously one population outperformed the others. So it can be assumed that the microorganisms in the MBR at the manufacturing facility are acclimated to the waste and biodegrade DMF and the other compounds at a faster rate then a non-acclimated population could. The implication of this is that populations can become acclimated to DMF and more efficiently remove BOD, and may become acclimated to partially oxidized PVP as well. If it is not necessary to complete oxidize PVP, the treatment costs would be much lower.

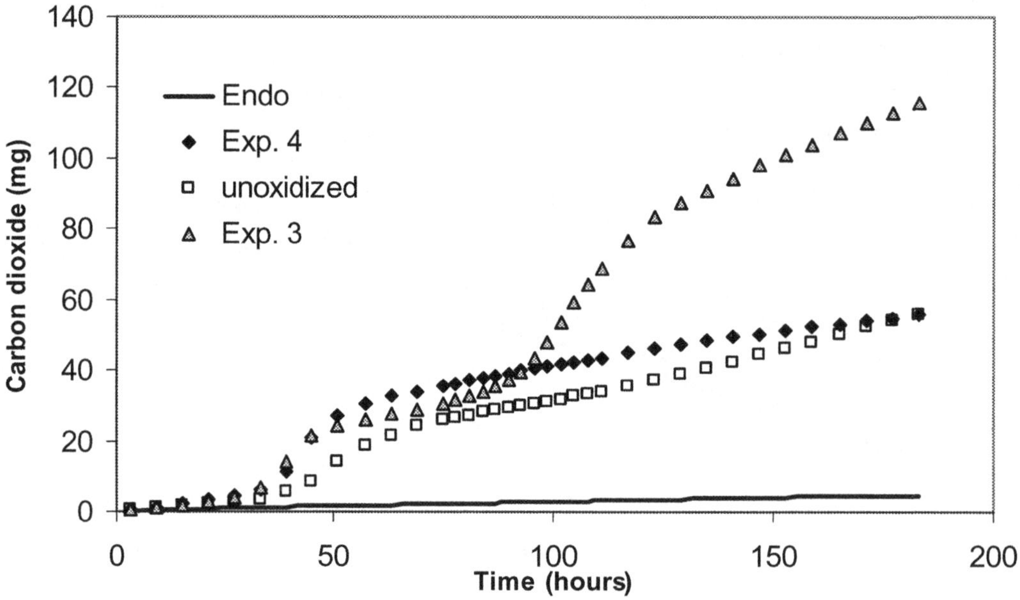

Figure 8-11. Long-Term Respirometric Test Results for Simulated Waste.

CHAPTER 9.0

CONCLUSIONS AND RECOMMENDATIONS

9.1 Conclusions

9.1.1 Chemical Oxidation

All three oxidation methods—ozonation, Fenton treatment, and UV/H_2O_2 treatment—accomplished oxidation of DCDE and MTBE to identifiable oxidation products. Ozone did not react with DCDE and MTBE directly but through the formation of hydroxyl radicals. Consequently, the process was slow and limited by the formation rate of the free radicals. Yet, oxidation of each mole of DCDE required 2.6–3.6 moles of ozone to decompose to free radicals. The range is typical and is also an indication that much higher ozone requirements will be observed if the wastewater contains additional organic compounds.

MTBE was found to be quite volatile. A substantial amount of MTBE was removed during ozonation by physical stripping, and as a result ozonation was not considered to be an effective method for MTBE removal.

Oxidation of DCDE and MTBE by Fenton reagents was successful, especially by stepwise addition of H_2O_2 and iron salts. Using relatively high concentrations of iron salts compared to H_2O_2 led to more efficient removal of DCDE and MTBE. However, this procedure also led to large amounts of sludge production and a requirement for excessive amounts of base for restoration of pH. This process requires optimization based on total cost of the treatment. Overall, ferric sulfate provided slightly higher efficiencies than ferric chloride as the iron source. However, using ferrous sulfate proved to be much more effective. For example, 3.3 moles of Fe (II) as ferrous sulfate and peroxide were needed per mole of MTBE oxidized. This stoichiometric factor for Fe (III) and peroxide was about 12 moles per mole of MTBE oxidized when ferric sulfate was used under the same conditions. For DCDE oxidation, about 25 moles of Fe (III) and peroxide were needed to oxidize one mole of DCDE.

Both DCDE and MTBE were easily oxidized by the UV/H_2O_2 method. There was minimal stripping of MTBE by this technique. Again, stepwise addition of peroxide increased the efficiency substantially and resulted in much less peroxide use for oxidation of the target chemicals. About 8.3 moles peroxide were needed to oxidize one mole of DCDE. This ratio was 6.8 moles of peroxide per mole of MTBE.

Oxidation by-products identified for each target chemical and for the three different oxidation methods were compared to each other after normalizing the data by presenting the molar carbon concentrations of the by-products as a function of percent oxidation of the parent compound. The product distribution for the three oxidation method showed that oxidation of DCDE progressed by different pathways. For example, formaldehyde formed primarily with ozone. Formic acid formed primarily for Fenton reagent and glycolaldehyde formed primarily with UV/H_2O_2. The majority of the identified reaction by-products were non-chlorinated aldehydes and carboxylic acids. However, all three oxidation methods resulted in about 55% dechlorination of DCDE when 90% of DCDE was oxidized. This indicates the presence of chlorinated organic by-products as well. Some of these products were identified.

Oxidation by-products of MTBE also showed differences among the three oxidation methods. Ozonation produced more tertiary butyl formate, acetone, and formaldehyde than the other processes. Fenton process, on the other hand, produced more tertiary butyl alcohol and formic acid. Acetic acid was observed only in UV/H_2O_2 oxidation.

9.2 Assessment of Biodegradation for Pre-oxidized DCDE

DCDE was shown to be non-biodegradable by the short-term and long-term respirometric tests. The short-term respirometric tests indicated that all three oxidation methods significantly increased the biodegradability of preoxidized DCDE solutions. The μ_{max} value increased from zero for non-oxidized DCDE to 0.1–0.2 mg oxygen consumption per liter per minute for the oxidized DCDE solutions. The differences among the oxidation methods were not significant. There was no significant difference between the acclimated and non-acclimated cultures either.

Long-term respirometric tests showed that after an initial period of about 70 hours for acclimated cultures and only 10 hours for non-acclimated cultures, Fenton reagent produced highly toxic by-products that have inhibited the metabolic activity of the microorganisms. The non-acclimated culture showed much more sensitivity to these products than the acclimated culture. DCDE samples oxidized by ozone and UV/H_2O_2 showed continued metabolic activity for 14 days. Ozone-treated samples exhibited significantly higher metabolic rate with non-acclimated cultures compared to the others. Furthermore, DCDE oxidized at higher oxidation levels before the respirometric tests exhibited higher metabolic rates than those oxidized at lower oxidation levels.

Mid-term respirometric tests indicated that higher oxidation levels produced higher concentrations of more readily biodegradable by-products, and that led to mineralization of a larger fraction of the organic matter through subsequent biodegradation. However, even after 90% oxidation of DCDE, only 25% of the oxidation by-products biodegraded within 6 hours as measured by TOC. No significant differences were observed among the oxidation methods. Both cultures produced similar results.

Inhibition studies indicated that ozonated and Fenton-treated samples resulted in 36%–72% inhibition of the respirometric rates of both acclimated and non-acclimated cultures. On the other hand, UV/H_2O_2-treated samples did not show any inhibition at any level of oxidation.

The SBR studies of DCDE oxidized samples showed that the reactors that received the oxidized DCDE almost always produced less removal of COD and TOC compared to the controls. However, the difference in COD and TOC removal efficiencies of these reactors and the controls got reduced with time, showing considerable acclimation of the organisms to the oxidation products. The fluctuations of SS, TOC, and COD in the reactors were observed to be due to introduction of large volumes of oxidized solutions, not so much large concentrations of the oxidized by-products. The only exception was the Fenton-treated DCDE solutions for which considerable reduction in efficiencies was observed in spite of accumulation of the suspended solids due to iron coagulation. Overall, the reactors initially biodegraded 50% to 70% of the oxidation by-products. However, after acclimation, these percentages increased to 80%–100%. No toxicity or inhibition of oxidation products was obvious, except for Fenton-treated solutions when they constituted 75% of the feed solution by volume. In partially-oxidized samples, the remaining DCDE was partially removed by adsorbing on biomass. Otherwise, DCDE did not biodegrade.

Overall, the short-term, long-term, and mid-term respirometric tests and the SBR studies showed that the oxidation by-products of DCDE were biodegradable. However, long-term tests indicated that oxidation by-products of Fenton treatment created toxicity after a certain initial time period. The inhibition studies showed that not only Fenton but ozone-treated solutions also created inhibition on acetate degrading biomass. None of the inhibition or toxicity of the oxidation by-products was obvious in the SBR tests. Furthermore, although only about 25% of the DCDE oxidation by-products were biodegradable according to the mid-term tests, the SBR tests showed that up to 90%–100% of the oxidation by- products can be biodegraded through acclimation to these compounds.

9.3 Assessment of Biodegradation for Pre-oxidized MTBE

All the biodegradation tests showed that MTBE was slowly biodegradable, and the biodegradability of MTBE solutions improved after chemical oxidation, i.e., chemical oxidation of MTBE resulted in formation of by-products that were more biodegradable than MTBE itself. In addition, increasing oxidation levels produced higher respiration rates for all three oxidation methods.

The maximum respiration rate observed for all three oxidation methods by the short-term respirometric test was about 0.15–0.2 mg/L-min. In long-term respirometric tests the acclimated culture exhibited a significantly higher respiration rate than the non-acclimated culture for all the three oxidation methods. Furthermore, significantly higher respiration rates were observed for the UV/H_2O_2-treated sample than those treated by ozone or Fenton reagent. No inhibition of Fenton-treated samples was obvious. The MTBE samples oxidized by 100% consumed much higher net oxygen and produced much higher net carbon dioxide levels than those oxidized by 75%. This indicates that the compounds produced at the 100% oxidation level are more biodegradable than those produced at 75% oxidation.

Mid-term respirometric tests indicated that about 57% of the oxidation by-products of UV/H_2O_2 treatment biodegraded within 24 hours as measured by TOC. Acclimated and non-acclimated cultures produced similar results.

Toxicity studies indicated no inhibition of the oxidized MTBE solutions. In fact, MTBE oxidation products caused negative inhibition, i.e., they served as secondary substrate. No major differences were observed between the acclimated and non-acclimated cultures.

The SBR studies of samples containing MTBE oxidation by-products showed that the reactors that received the oxidation by-products produced less removal of COD and TOC compared to the controls. Based on the results, it was estimated that only about 50% of the oxidation products biodegraded under the experimental conditions. This result agrees with the mid-term respirometric results that about 57% of the by-products were biodegradable. There was no further acclimation over a 12-day period.

9.4 Assessment of Biodegradation for Pre-oxidized Groundwater and Industrial Wastewater

Groundwater containing various petroleum contaminants in addition to MTBE was successfully oxidized to remove MTBE. However, oxidant requirement was much higher than pure MTBE solutions. The results showed significantly improved biodegradation after chemical oxidation based on long-term respirometric studies. The SBR tests showed that about 63% of the oxidation by-products of groundwater were biodegradable.

It was possible to oxidize PVP, which is a non-biodegradable organic compound in a waste matrix that had much larger concentrations of other organic compounds. However, it was estimated that only about 2.5% of the oxidant was used on the targeted PVP; the rest was consumed or scavenged by other compounds. After oxidation, the PVP solutions showed highly increased biodegradability as measured by CO_2 formation. Similarly, simulated waste containing oxidized PVP also exhibited improved biodegradation even after 45% oxidation.

9.5 Evaluation of the Biodegradability and Toxicity Tests

Overall, five different tests were employed in this study:

- Short-term respirometry
- Long-term respirometry
- Mid-term biodegradation test
- Inhibition test
- SBR test

Each of the tests provided valuable information about the effectiveness of chemical oxidation to improve the biodegradability of the solutions containing DCDE or MTBE.

The short-term respirometric test is simple to execute. It takes only about 30 minutes to get some results. In fact, all the oxidized samples of DCDE and MTBE at every level of oxidation showed an immediate increase in net oxygen consumption, indicating the formation of biodegradable compounds from chemical oxidation. This test is a valuable tool for preliminary screening. However, extracting any other information from this test, e.g., comparison of behavior of acclimated versus non-acclimated cultures, obtaining Monod constants, trying to understand the inhibition or toxicity of the oxidation by-products, will be difficult and most likely misleading due to many limitations of the test.

The long-term test is more tedious requires a more sophisticated instrument. However, it may provide information about possible acclimation of the culture to the oxidation by-products and possible negative response of the culture to toxic or inhibitory compounds. For both DCDE and MTBE, the long-term test showed that increasing the oxidation level increased the biodegradability of the oxidation by-products. In other words, higher oxidation levels produced by-products that were more biodegradable. Furthermore, all three oxidation processes provided very positive results, except Fenton treatment of DCDE solutions.

The mid-term tests were designed to provide information about the extent of biodegradation of the oxidation by-products within 6 hours—a typical residence time in a treatment plant—in terms of TOC and COD removal. This test showed 25% biodegradation for DCDE oxidation by-products and about 57% biodegradation for MTBE oxidation by-products. This test did not allow the acclimation of the cultures to the oxidation by-products and, as a result, showed the minimum expected biodegradability after chemical oxidation. For example, the SBR results with oxidized DCDE solutions indicated that the by-products of chemical oxidation could be biodegradable by 90–100% after going through an acclimation period.

The toxicity tests were conducted to evaluate if the chemical oxidation by-products can cause inhibition of the microorganisms that use a readily biodegradable substrate. This test mimicked a municipal activated sludge process that received chemically oxidized industrial wastewater. The results of the test, with acetic acid as the primary substrate and oxidation by-products of DCDE or MTBE, showed that oxidation by-products of MTBE were not toxic to the biomass at all. Oxidation by-products resulting from UV/H_2O_2 treatment of DCDE did not show any toxicity either. However, oxidation by-products of ozonation and Fenton treatment of DCDE caused considerable inhibition. Long-term tests confirmed the toxicity of the products of Fenton-treatment of DCDE but not of the ozonation. However, the SBR tests showed no toxicity for the DCDE oxidation by-products. This was probably because the relatively large biomass concentrations used in SBRs (2,000–3,000 mg/L) compared to the mid-term tests (400 mg/L). In addition, the inhibition tests used acetic acid as the main substrate, whereas the SBRs used a more complex feed that is more representative of municipal wastewater.

The SBR test was quite successful in providing most of the needed information in one single test. The information obtained through SBR included:

- Whether the chemical oxidation by-products were biodegradable under the operational conditions of a typical activated sludge plant;
- What percent of the oxidation by-products were biodegradable;
- Whether the chemical oxidation by-products caused any inhibition on the biomass; and
- Whether the biomass showed any acclimation during the treatment to the oxidation by-products.

However, it was difficult to understand the fate of the chemicals in SBR tests alone. For example, short-term, long-term, and mid-term tests all showed that non-oxidized DCDE did not biodegrade at all. Yet, in the SBR studies about 40% of 50 mg/L of non-oxidized DCDE introduced with the feed did not show up in the effluent. It is obvious that DCDE got adsorbed on the biomass since it did not strip under vigorous mixing conditions, and it is a chemical with high affinity for adsorption. On the other hand, non-oxidized MTBE is believed to be stripped during the SBR test. Under similar mixing conditions, MTBE was shown to be easily stripped from the water. Furthermore, short- and long-term tests showed that MTBE biodegradation is much slower than is its rate of volatilization under SBR conditions.

The SBR tests indicated considerable acclimation of the biomass to the oxidation by-products of DCDE, and as a result, up to 100% biodegradability of the by-products was observed after the biomass got acclimated. On the other hand, the biomass did not show any acclimation to the MTBE oxidation by-products. The toxicity studies with acetate degrading biomass showed significant inhibition of DCDE oxidation by-products, especially with ozone and Fenton treatment. Yet, no inhibition of MTBE oxidation by-products was observed. Furthermore, the mid-term tests indicated that only about 25% of the oxidation products were biodegradable under the experimental conditions. This value was 57% for MTBE oxidation by-products. When the results of the SBR, the toxicity, and the mid-term tests are all considered together, the following observation seems to emerge: If the oxidation by-products are biodegradable yet toxic, mid-term test may indicate low biodegradability. Yet the activated sludge under the operational conditions of typical treatment plants get acclimated to the toxicity of the by-products and as a result can biodegrade all of the oxidation by-products. However, if the oxidation products are slowly biodegrading, e.g., TBA from MTBE, although they may not be toxic, activated sludge operation may not improve the level of biodegradation shown already by the mid-term tests because the oxidation products are slow to degrade.

Because there were no observed differences based on the SBR tests among the oxidation methods in terms of biodegradability or toxicity of the oxidation by-products, no attempt was made to further evaluate the biodegradability and toxicity of the oxidation products. It should be noted the slight reduction in the efficiency of the reactor that received Fenton-treated solution was obvious only when the volume contribution of the oxidized DCDE to the feed was raised to 75%. So, although it is clear that Fenton treatment of DCDE may be producing toxic chemicals that might potentially create upsetting conditions in reactors, they can be managed if their volume contribution, i.e., the concentration of the toxic by-products can be maintained at low levels.

Another conclusion of this study is that the acclimation of the biomass to be used in the tests is not useful if the target chemical is not biodegradable, for example, DCDE. On the other hand, if the target chemical is slightly biodegradable, e.g., MTBE, acclimation of biomass to the target chemical is beneficial, especially if the target chemical is not chemically oxidized by 100%. However, if the target chemical needs to be oxidized by 100%, i.e., it is non-biodegradable or toxic, then acclimating the biomass to the target chemical does not provide any additional benefits. The biomass should instead be acclimated to the oxidation by-products.

9.6 Recommendations

Chemical oxidation methods are recommended to remove various types of non-biodegradable or toxic chemicals present from industrial wastewaters or contaminated groundwater. In limited cases, the removal of the targeted contaminant(s) may allow the discharge of the wastewater directly to the environment. However, more frequently the chemically oxidized wastewater will need further treatment for removal of the oxidation by-products as well as other biodegradable compounds. This can best be accomplished by integration of the preoxidation process by a biological treatment system.

The type of chemical oxidation method to be applied before biotreatment needs to be decided based on the overall economics. The majority of the target chemicals will be more readily oxidizable by hydroxyl radical rather than ozone molecule. Although ozone can be used as a source of hydroxyl radical, it is usually more costly in terms of capital investment than other oxidation methods. Furthermore, volatile organics may be stripped during mixing and stirring needed in some process applications, e.g., ozonation. This obviously is not desired, unless there is a mechanism by which the stripped organic compounds are collected and treated further in the gas phase. Fenton application is much less capital intensive than the UV/H_2O_2 process applied in this research. However, more chemicals are needed per mole of target organic compound. Furthermore, handling and disposal of sludge could be difficult. Real wastewaters or contaminated groundwaters pose additional challenges for chemical oxidation processes, including high turbidity, alkalinity, and more importantly, the presence of highly concentrated non-targeted organic compounds. The UV/H_2O_2 process is very sensitive to the turbidity. The presence of high alkalinity and high concentrations of non-targeted organic compounds pose a challenge for all oxidation techniques that rely on hydroxyl radical formation. The required amounts of chemicals and energy needed for treatment of wastewaters therefore far exceeds the requirements estimated based on pure solutions of the target chemicals.

If total removal of the target compound(s) seems to be economically challenging, partial oxidation should be considered especially for target compounds that are slowly biodegradable and nontoxic.

For future studies, it is recommended to use short-term respirometric tests for screening, followed by SBR tests conducted under conditions more representative of typical activated sludge plants for a more comprehensive evaluation of the pretreatment techniques. The procedures described in this study for evaluation of the SBR test results are recommended for future studies as well. By focusing on these two tests only, the testing time can be shortened considerably.

Appendix A

Table A-1. The Originally Proposed Test Chemicals and Their Selected Properties.

Compounds	General		Physical Properties[1]			Biodegradability[2]		
	Formula	M.W.	Solubility in water (mg/L)	$logK_{ow}$	$logK_H$	TOD (g/g)	BOD_5 (g/g)	BOD_5/TOD (%)
Chlorinated Aromatics								
1. Chlorobenzene	C_6H_5Cl	112.56	200	2.92	0.55	2.00	0.03	1.5
2. 1,4-Dichlorobenzene	$C_6H_4Cl_2$	147.01	123	3.38	0.35	1.41	0.11	7.8
3. Hexachlorobenzene	C_6Cl_6	284.77	0.006	5.5	0.18	0.51	0	0
4. Pentachlorophenol	C_6HOCl_5	266.34	14	5.12	-2.63	1.00	0	0
Aromatic Hydrocarbons								
5. Ethyl benzene	C_8H_{10}	106.07	152	3.15	0.9	3.17	0.92	29
6. 1,2,4-Trimethylbenzene	C_9H_{12}	120.2	60	3.63	0.76	3.20	0	0
7. 1,3,5-Trimethylbenzene	C_9H_{12}	120.2	57	3.42	0.92	3.20	0	0
Chlorinated Aliphatics								
8. Dichloroethylene	$C_2H_2Cl_2$	96.95	800	2.06	-	0.66	0.05	0
9. Dichloropropylene	$C_3H_6Cl_2$	112.99	3,000	1.15	-2.55	1.13	0.19	16
10. Methylene chloride	CH_2Cl_2	84.93	13,200	1.25	0.41	0.38	0	0
Ethers								
11. Dichlorodiethyl ether	$C_4H_8OCl_2$	143.02	10,200	1.29	-	1.12	0	0
12. Glycol ether	$C_4H_{10}O_3$	106.14	10,000	0.89	-	1.25	0.12	7.9
13. Methyl tert-butyl ether	$C_5H_{12}O$	88.2	48,000	1.3	-	2.73	0.55	20

TOD: Theoretical oxygen demand

$logK_{ow}$: log octanol/water partition coefficient

$logK_H$: log Henry's Law constant

(1) Schwarzenbach et al., 1993

(2) Pitter and Chudoba, 1990

Although the original number of chemicals proposed for this research was 13, as listed in Table A-1., the Project Subcommittee and the project team decided to limit the number of the chemicals to two, namely DCDE and MTBE, in order to emphasize:

- the understanding of the fundamentals of the improvement of biodegradability upon oxidation; and
- the development of a procedure, which can be tested on a smaller number of chemical compounds.

Nevertheless, quite a number of experiments were conducted on some of the chemicals listed in Table A-1. The following conclusions were derived based on these experiments:

1. Glycol ether (diethylene glycol, DEG) was found biodegradable by non-acclimated activated sludge, requiring no prior chemical oxidation. The results of long-term respirometric tests are presented in Figure A-1 as a function of DEG concentration.

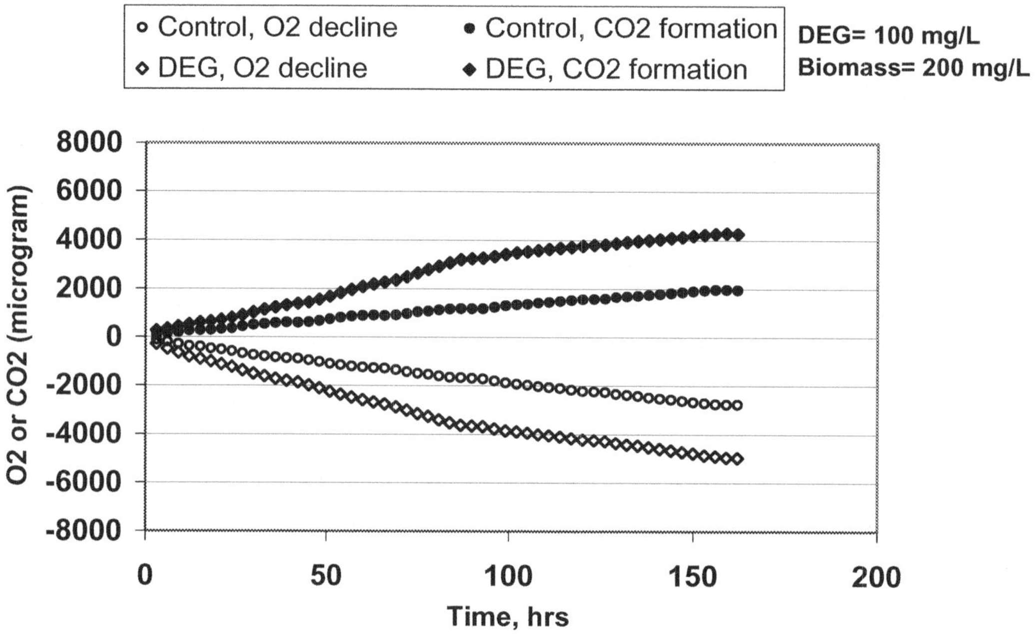

Figure A-1. Long-Term Respirometric Results for DEG.

2. Long-term respirometric tests on 1,2,4-trimethylbenzene (TMB), ethyl benzene (EB), and methylene chloride (MC) with non-acclimated activated sludge also showed higher oxygen uptake and carbon dioxide production compared to the controls (Figures A-2 through A-4). Chemicals TMB, EB and MC are relatively volatile compounds, and the GC analysis after the respirometric runs revealed no traces of these chemicals in solution. Yet, an enhanced oxygen uptake and carbon dioxide production was observed in the respirometric studies. This could be due the microorganisms being able to start utilizing these chemicals immediately.

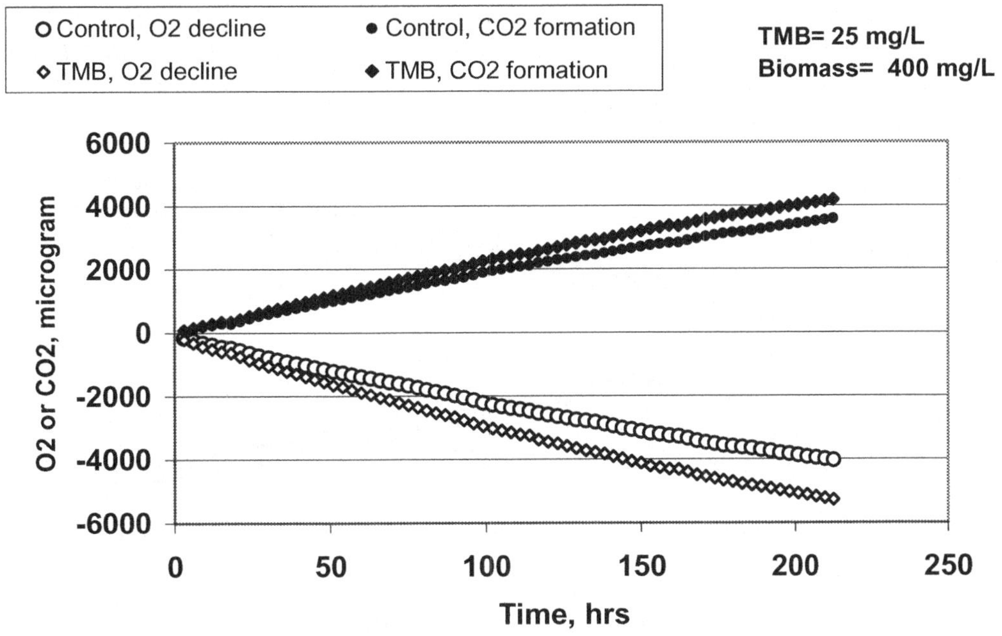

Figure A-2. Long-Term Respirometric Results for TMB.

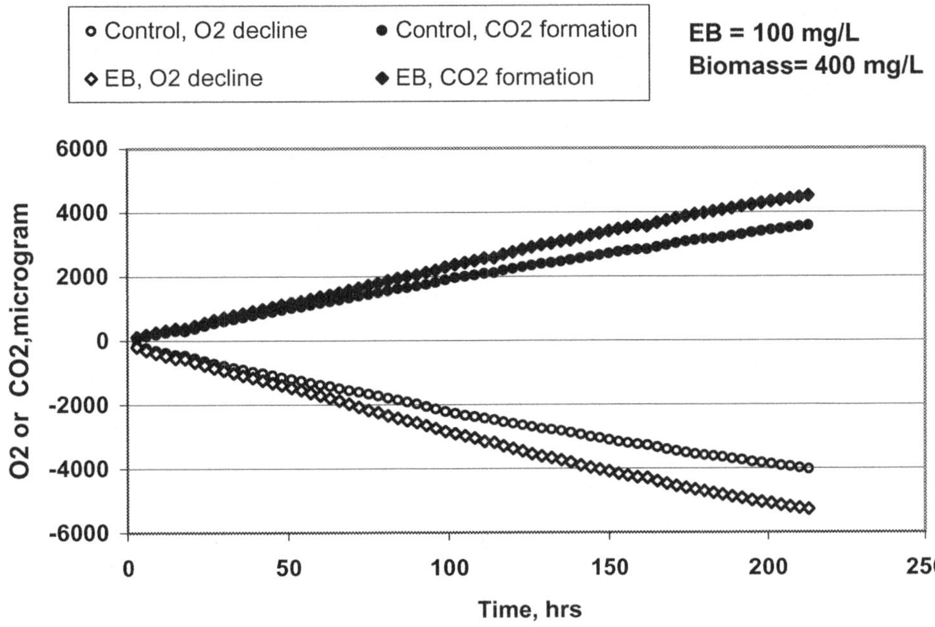

Figure A-3. Long-Term Respirometric Results for EB.

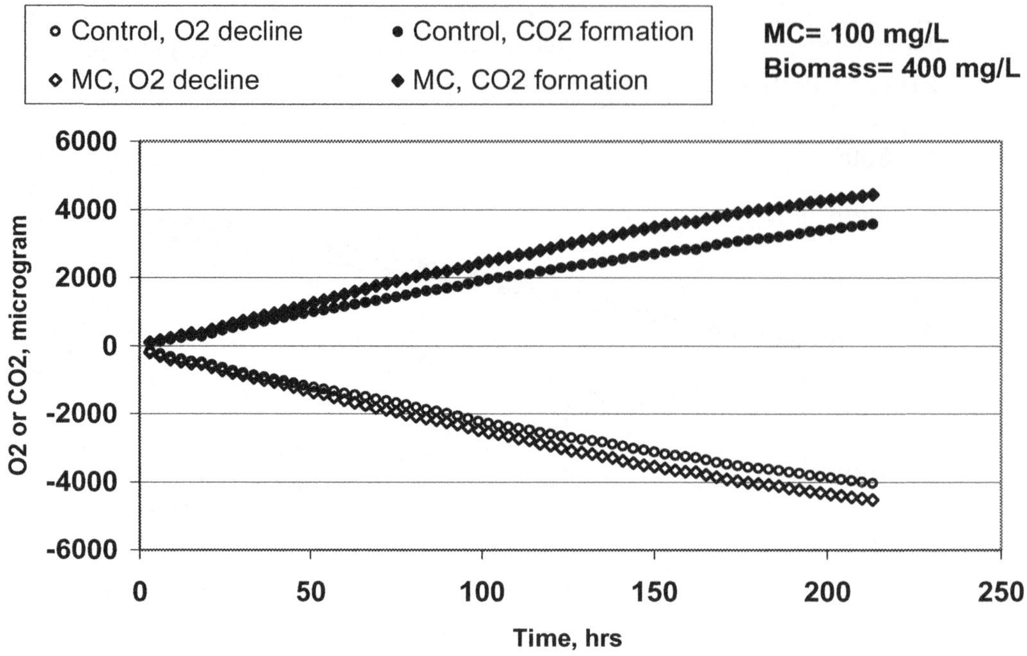

Figure A-4. Long-Term Respirometric Results for MC.

3. Long-term respirometric tests on dichloro propylene (VDC) showed that the oxygen uptake and carbon dioxide production were lower than those for the endogenous controls (see Figure A-5). This chemical might be toxic to the biomass. Even though VDC is a very volatile compound and no trace of VDC was found in solution at the end of the respirometric run, it seems that it can still exert considerable toxicity during the beginning of the test.

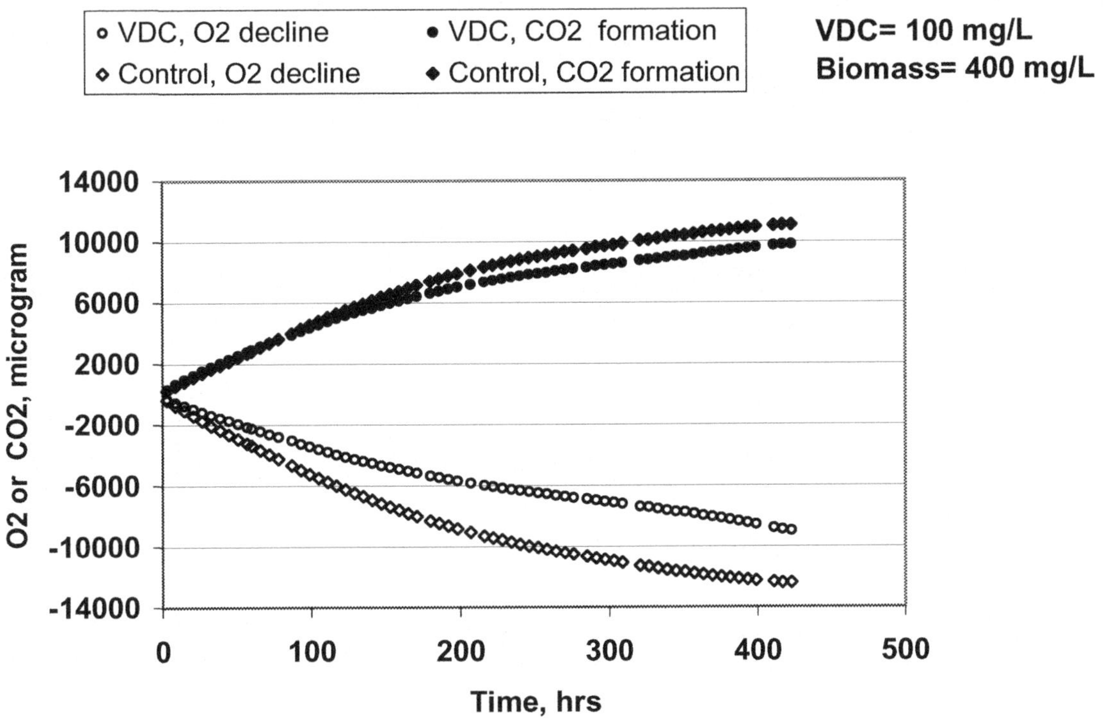

Figure A-5. Long-Term Respirometric Results for VDC.

4. The oxygen uptake and carbon dioxide production for 1,3,5-trimethyl benzene (MST), dichloro propylene (PDC) and chlorobenzene (CB) were comparable to those of the controls, i.e., there was no difference between the tests and controls with non-acclimated activated sludge culture (Figures not shown). This group of chemicals consists of highly volatile compounds. Therefore, one can fairly anticipate the chemicals will escape from solution and be adsorbed onto the activated carbon trap used on the pump sampling line of the instrument. In fact, analysis for these organic chemicals after the respirometric runs revealed no traces of the chemicals in solution. Accordingly, the respirometric rates default to the endogenous rates. Hence, one can simple conclude that biodegradability assessment by the respirometer in use will not work for the volatile chemicals.

5. Inhibition tests indicated a high level of inhibition of PCP on the metabolic rates of the non-activated activated sludge. The results are presented in Figure A-6 as a function of the PCP concentration presented as a fraction of water solubility limit of 14 mg/L.

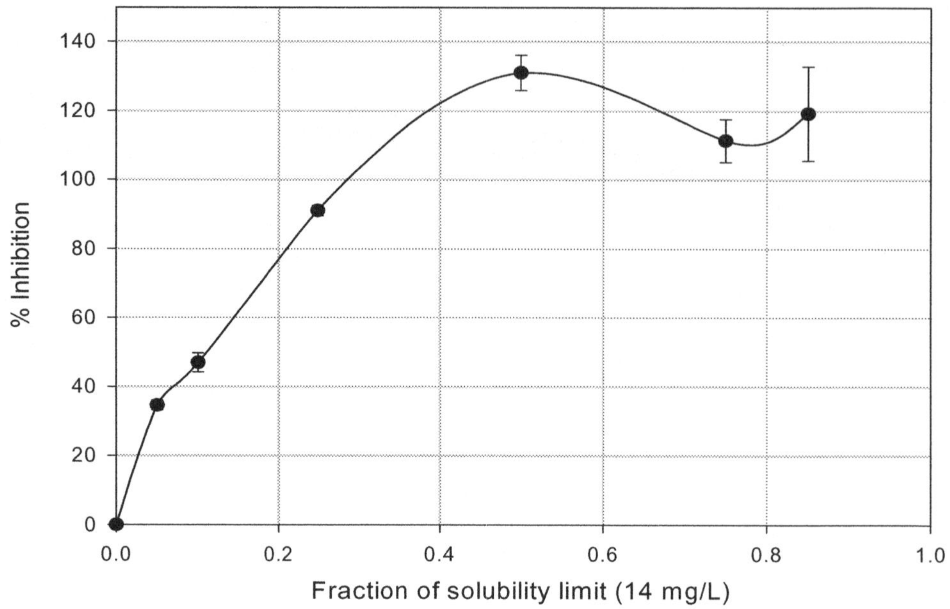

Figure A-6. Results of Inhibition Tests on PCP.

6. The accuracy of the respirometer for measurements of oxygen and carbon dioxide was verified using a mixture of glucose and glutamic acid. Two concentrations of the mixture were used in duplicate bottles. The concentration utilized were 50 mg/L + 50 mg/L and 100 mg/L + 100 mg/L glucose + glutamic acid, respectively.

The results of these studies for the 50/50 mixture are shown in Figure A-7. For both solutions, the respirometer tracked the degradation of this easily biodegradable mixture very well both in accuracy and reproducibility. The total oxygen consumption after 66 hours of operation matched well with theoretical biochemical oxygen demand (TBOD) for this mixture. The points of TBOD were represented on the figure by a full circle. Similarly, the carbon dioxide production as measured by the respirometer matched with the anticipated carbon dioxide generation according to stoichiometric calculation. The anticipated carbon dioxide production values were represented on the graphs by a full triangle.

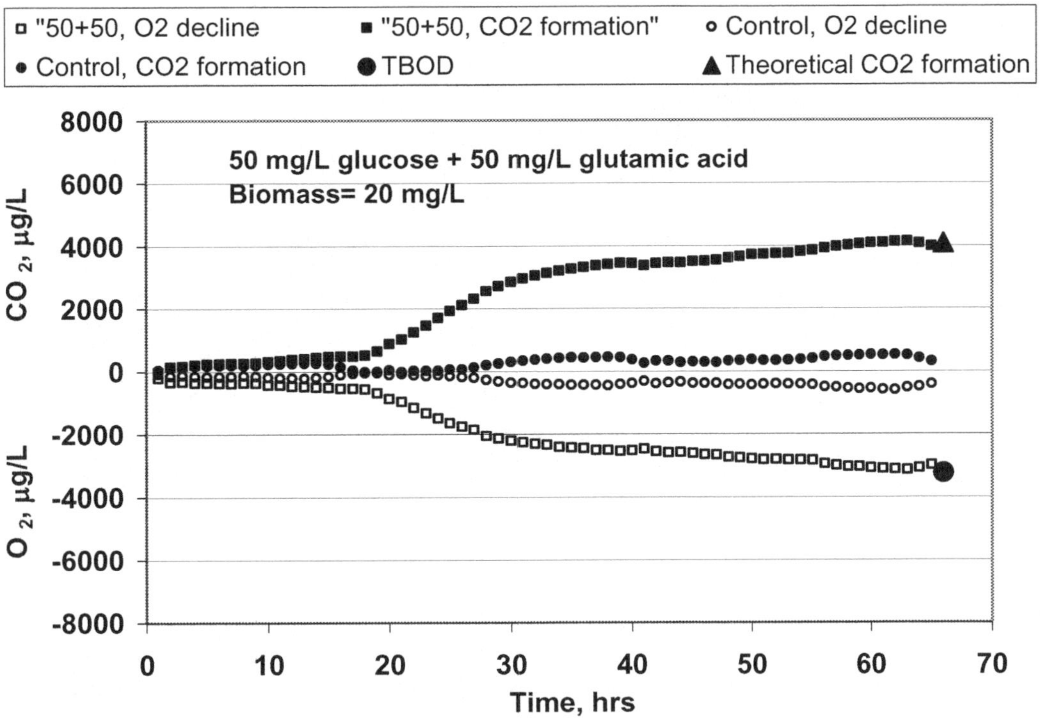

Figure A-7. Respirometric Studies for Mixture of Glucose and Glutamic Acid.

7. The following compounds were found to volatilize rapidly under the conditions applied for ozonation at 0.5 L/min gas flow rate: EB, TMB, MST, VDC, PDC, and MC. Similarly, these chemicals were found to be lost considerably during Fenton treatment to volatilization, adsorption on iron oxide precipitate, and sample filtration for separation of the precipitate from the solution. The results of these tests for aeration (mimicking ozonation) and Fenton treatment (without hydrogen peroxide) are presented in Figures A-8 and A-9, respectively.

According to the results presented in these figures, these chemicals can be removed by 35–75% over the reaction period, depending on the chemical, due to stripping during ozonation under the experimental conditions tested in this study. For comparison, it should be noted that the loss of MTBE under the same experimental conditions was 46%. The loss of the chemicals during handling for Fenton treatment due to stripping, adsorption, and filtration ranged from 40% for PDC to 92% for MST.

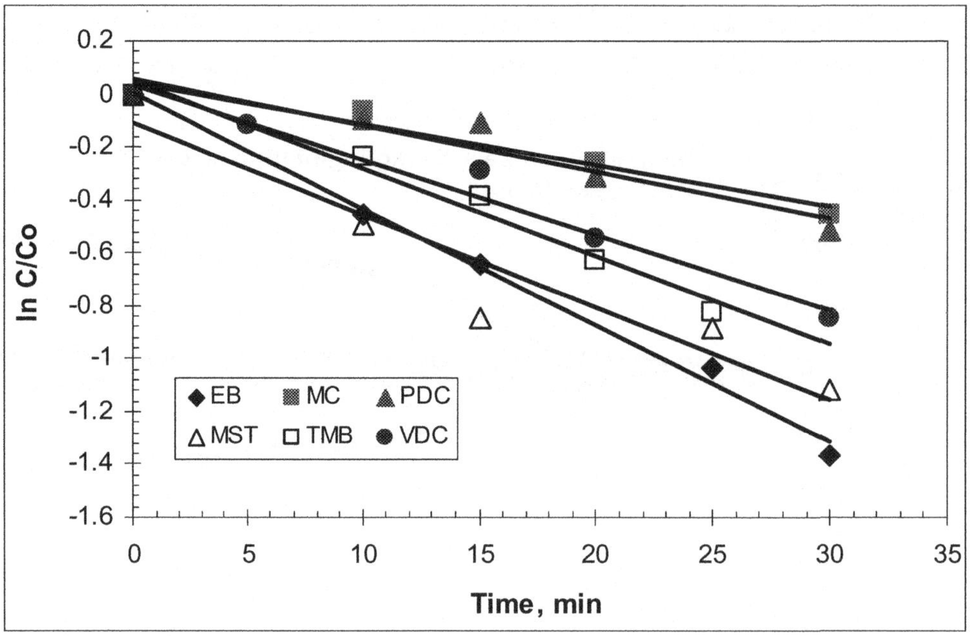

Figure A-8. Volatilization of Chemicals During Aeration (the same conditions as ozonation without ozone).

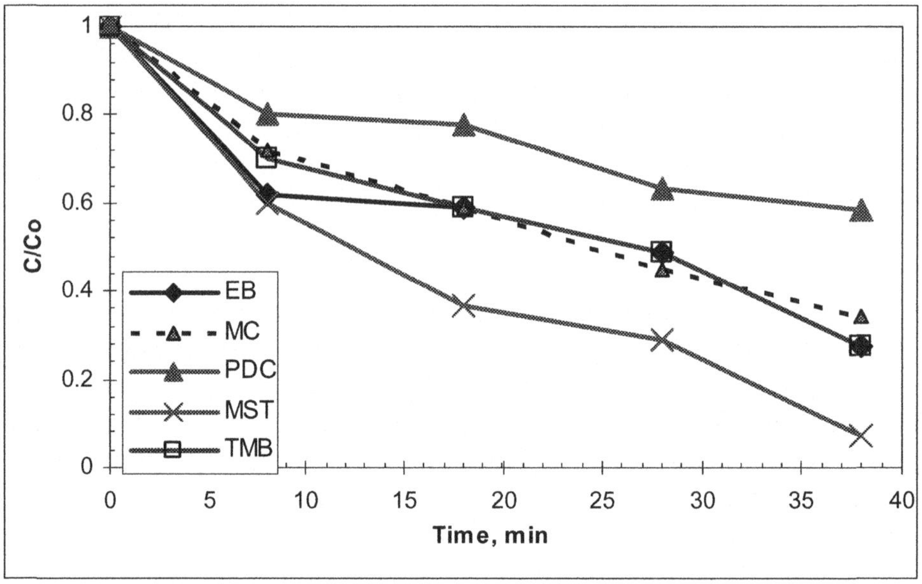

Figure A-9. Removal of VOCs During Fenton Treatment, Except Without Hydrogen Peroxide.

APPENDIX B

Theses Resulting from This Research

Milica Kaludjerski "Enhancement of Biodegradation of DCDE by Pre-Oxidation", M.S. Thesis, San Diego State University, 2001.

Melissa Valdovinos "Inhibition and Toxicity Assessment of Pre-oxidized Dichloro Diethyl Ether", M.S. Thesis, San Diego State University, 2002.

Alex Christensen "Utilizing SBRs to Determine Biodegradability of by-products Obtained from DCDE and MTBE after Oxidation", M.S. Thesis, San Diego State University, 2002.

Publications Resulting from This Research

"Assessment of Enhancement in Biodegradation of Dichlorodiethyl Ether (DCDE) by Pre-oxidation" by M. Kaludjerski and M.D. Gurol, *Water Research,* in press, 2004.

"Oxidation of Diethylene Glycol (DEG) with Ozone and Modified Fenton" by T. Ertas and M.D. Gurol, *Chemosphere,* vol. 47, 3, pp.293-301, 2002.

"Improvement of Biodegradability of Persistent Organic Compounds by Chemical Oxidation", by M.D. Gurol, *Proceedings of Eight Annual Industrial Waste Conference*, 2002.

"Enhancement of Biodegradation of Dichlorodiethyl Ether by Pre-Oxidation" *In-Situ Aeration and Aerobic Remediation,* M. Kaludjerski, F. Buyuksonmez, T. Ertas and M. Gurol, Ed. by Leeson, et al., Batelle Press, 6(10), pp. 301-309, 2001.

Presentations Resulting from This Research

"Improvement of Biodegradability of Persistent Organic Compounds by Chemical Oxidation" by M. Gurol, 8th Annual Industrial Wastes, WEF, Atlantic City, NJ, August 11-14, 2002.

"MTBE Removal by Chemical Oxidation" by M.D. Gurol, C. Arias, and T. Ertas. AWWA Annual Conference in Washington, D.C. June 2001.

"Pre-treatment of Persistent Organic Compounds by Chemical Oxidation for Improvement of Biodegradability" by M. Gurol, et al., VIIth International Conference on Advanced Oxidation Technologies for Water and Air Remediation Niagara Falls, Ontario, Canada, June 2001.

"Ozonation of Dichloroethylene (DCDE)" by T. Ertas and M. Gurol, International Ozone Association 2001 Pan American Group Conference, Newport Beach, CA, May 5-9, 2001.

"Ozonation and Stripping of MTBE" by T. Ertas and M. Gurol, The 11th Annual West Coast Conference on Contaminated Soils, Sediments and Water, San Diego, CA, March 19-22, 2001.

"Enhancement of Biodegradation of DCDE by Pre-Oxidation" by Kaludjerski, et al., Sixth International Symposium on In Situ and On-Site Bioremediation, San Diego, CA, June 2001.

“Determination of Biodegradation Kinetic Constants of Pre-Oxidized Solutions” by Buyuksonmez, et al., Sixth International Symposium on In Situ and On-Site Bioremediation, San Diego, CA, June 2001.

“Effect of Pre-Oxidation on the Inhibition of Activated Sludge Activity of DCDE” by Valdovinos, et al., Sixth International Symposium on In Situ and On-Site Bioremediation, San Diego, CA, June 2001.

“Protocol for the Measurement of Inhibition of Xenobiotic Compounds on Activated Sludge” by Buyuksonmez, et al., Sixth International Symposium on In Situ and On-Site Bioremediation, San Diego, CA, June 2001.

“Respirometric Assessment of the Degradability of MTBE and other Ethers” by Badriyha, et al., Sixth International Symposium on In Situ and On-Site Bioremediation, San Diego, CA, June 2001.

“The Use of Ozone, UV and Fenton-based Processes in Water and Wastewater Treatment” by M. Gurol, Department of Environmental Sciences, California Institute of Technology, Nov. 1, 2000.

Papers in Preparation

"Inhibition and Toxicity Assessment of Pre-oxidized Dichloro Diethyl Ether” by M. Valdovinos and M. Gurol, to be submitted to *Water Environment Research.*

“Utilizing SBRs to Determine Biodegradability of Oxidation By-products of DCDE and MTBE” by A. Christensen and M. Gurol, to be submitted to *Water Environment Research*.

“A Critical Review of the Biodegradability Tests” by M.Gurol, to be submitted to *Water Environment Research.*

“A Comparison of Chemical Oxidation Processes for Removal of the Parent and the Oxidation Products of DCDE” by M.Gurol and T. Ertas, to be submitted to *Water Research*.

“A Comparison of Chemical Oxidation Processes for Removal of the Parent and the Oxidation Products of MTBE” by M.Gurol, T. Ertas, and T. Garoma, to be submitted to *Water Research.*

References

Adams, C.D., R.A. Cozzens, and B.J. Kim (1997) Effects of ozonation on the biodegradability of substituted phenols, *Water Research*, 31 (10): 2655–2663.

Adams, C.D. and J.J. Kuzhikannil (2000) Effect of UV/H_2O_2 preoxidation on aerobic biodegradability of quaternary amine surfactants, *Water Research*, 34 (2): 668–672.

Amiri, A.S. O_3/H_2O_2 treatment of methyl-*tert*-butyl ether (MTBE) in contaminated waters. *Water Res.*, 35 (15):3706–3714.

ATSDR, Agency for Toxic Substances and Disease Registry (1989) Toxicological profile for bis (2-chloroethyl) ether, ATSDR/TP-89/02.

Bader, H. and J. Hoigne (1981) Determination of ozone in water by the indigo method. *Water Res.,* 15: 449–456.

Baillod, C.R., Lamparter, R.A., and Barna, B.A. (1985) Wet oxidation for industrial waste treatment, *Chemical Engineering Progress,* 81 (3): 52–56.

Baker, R. and J.L. Warren (1992) Generation of hazardous waste in the United States, *Hazardous Waste & Hazardous Materials*, 9(1): 19–35.

Barb, W.G., J.H. Baxendale, P. George, and K.R. Hargrave (1951) Reactions of ferrous and ferric ions with hydrogen peroxide, *Trans. Faraday Soc.* 47: 462–500.

Baxendale, J.H. and J.A. Wilson (1956) The photolysis of hydrogen peroxide at high light intensities. *Trans. Faraday Soc.* 53: 344.

Beltran, F.J., J.F. Garcia-Araya, and P.M. Alvarez (1999) Wine distillery wastewater degradation. 1. Oxidative treatment using ozone and its effect on the wastewater biodegradability, *Journal of Agriculture and Food Chemistry*, 47: 3911–3918.

Beschkov, V., G. Bardarska, H. Gulyas, and I. Sekoulov (1997) Degradation of triethylene glycol dimethyl ether by ozonation combined with UV irradiation or hydrogen peroxide addition, *Water Sci. Technol.*, 36 (2–3): 131–138.

Bolduc, L. and W.A. Anderson (1997). Enhancement of biodegradability of model wastewater containing recalcitrant or inhibitory chemical compounds by photo catalytic pre-oxidation, *Biodegradation*, 8: 237–249.

Boltz, D.F. and J.A. Howell (1978). Colorimetric determination of nonmetals. John Wiley and Sons, New York.

Bossmann, S., E.G. Oliveros, S. Siegwart, E.P., Dahlen, L. Payawan, M. Straub, M. Warner, and A.M. Braun (1998). New evidence against hydroxyl radicals as reactive intermediates in the thermal and photochemically enhanced Fenton reactions. *J. Phys. Chem. A.* 102, 5542–550.

Bowers, A.R., P. Gaddipati, W.W. Eckenfelder, Jr., and R.M. Monsen (1989) Treatment of toxic and refractory wastewaters with hydrogen peroxide. *Water Sci. Tech.* 21: 477–486.

Burbano, A., D.D. Dionysiou, T.L. Richardson, and M. Suidan (2002) Degradation of MTBE intermediates using Fenton's reagent *J. Env. Eng.* 128(9): 799–805.

Buxton, G.V., C.L. Greenstock, W.P. Helman, and A.B. Ross (1988) Critical review of rate constants for reactions of hydrated electrons, hydrogen atoms and hydroxyl radicals in aqueous solution. *J. Phys. Chem. Ref. Data*, **17**, 513–886.

Can, Z. and M.D. Gurol "Interaction of Ozone with Formic Acid" Submitted to *Water Research,* January 2003.

Chamarro, E., A. Marco, and S. Esplugas (2001) Use of Fenton reagent to improve organic chemical biodegradability, *Water Res.*, 35 (4): 1047–1051.

Chang, P. and T. Young (2000) Kinetics of methyl *tert*-butyl ether degradation and by-product formation during UV/hydrogen peroxide water treatment, *Water Res.* 34(8): 2233–2240.

Christensen, A. (2002) Utilizing Sequential Batch Reactors (SBRs) to determine biodegradability of by-products obtained from dichlorodiethyl ether (DCDE) and methyl *tert*-butyl ether (MTBE) after oxidation. M.S. Thesis, San Diego State University.

Copa, W. and W.B. Gitchel (1988) Wet Oxidatio. In H.M. Freeman (ed) *Standard Handbook of hazardous waste treatment and disposal,*. Mc-Graw Hill.

Dietrich, M.J., T.L. Randall, and P.J.Canney (1985). Wet air oxidation of hazardous organics in wastewater, *Environ. Progress* 4(3): 171–177.

Djongalic, J. and J. Levec (1998) Does the catalytic wet oxidation yield products more amenable to biodegradation, *Applied Catalysis B: Environ.* L 1: L1–L5.

Ertas, T. and M.D. Gurol (2002). Oxidation of diethylene glycol (DEG) with ozone and modified Fenton. *Chemosphere*, 47 (3): 293–301.

Fisher, 2000. http://www.fishersci.ca /msds.nsf/96cb2019dad1311a85256670001d92b9 /8b657f2b9e292cfa852566f100053449. OpenDocument. Retrieved December 2000.

Gallard, H. and T. De Laat (2001) Kinetics of oxidation of chlorobenzenes and phenyl-ureas by Fe(II)/H2O2 and Fe(III)/H2O2. Evidence of reduction and oxidation reactions of intermediates by Fe(II) or Fe(III). *Chemosphere* 42: 405–413.

Gilbert, E. (1987) Biodegradability of ozonation products as a function of COD and DOC elimination by example of substituted aromatic substances. *Water Res.* 21 (10): 1273–1278.

Grady, C.P.L., Jr. (1985) Biodegradation: Its measurement and microbiological basis" *Biotechnology and Bioengineering*, 27: 660.

Gurol, M.D. (1985) Factors controlling the removal of organic pollutants in ozone reactors, *Journal AWWA*, 77(8): 55–60.

Gurol, M.D. (1993) Destruction of organic contaminants by non-thermal chemical oxidation processes, Invited presentation at the Association of Environmental Engineering Professors Research Opportunities Conference, Univ. of Michigan, Ann Arbor.

Gurol, M.D. and A. Akata (1996) Kinetics of ozone photolysis in aqueous solution, *AIChE Journal*, 42(11): 3283.

Gurol, M.D., C. Arias, and T. Ertas. (2001) MTBE Removal by Chemical Oxidation. Paper presented at the AWWA Annual Conference in Washington, D.C. June 2001.

Gurol, M.D. and C.H. Liao (1991) "Modeling of H2O2/UV Oxidation Process for Water Treatment in a Continuous Stirred Tank Reactor," *Proceedings of the Annual Meeting of AWWA*, pp. 857–873, Philadelphia, PA.

Gurol, M.D. and S. Nekouinaini (1984) "Kinetic behavior of ozone in aqueous solutions of substituted phenols, *Indust. Eng. Chem. Fundamentals,* 23(l): 54–60.

Gurol, M.D. and J.X. Ravikumar (1992) Enhanced biodegradation of pentachlorophenol and trichloroethylene in soil by chemical oxidation, *Proc. In-Situ Treatment of Contaminated Soil and Water*, Sponsored by Air & Waste Management Association and the U.S. EPA, Cincinnati, OH, pp. 245–256.

Hao, O.J., K.K. Phull, and J.M. Chen (1994). Wet oxidation of TNT red water and bacterial toxicity of treated waste, *Water Res.*, 28(2) 283.

Henze, M. (1992) Characterization of wastewater for modeling of activated sludge processes, *Water Sci. & Technol.*, 25(6): 1–15.

Hoffman, M.R. (1993) Chemical effects of collapsing cavitation bubbles and pulsed-power plasma discharges, *Proc. of First Intern. EPRI/NSF Symposium on Advanced Oxidation.* vol. 2.

Hoigne, J. and H. Bader (1983) Rate constants of reactions of ozone with organic and inorganic compounds in water, *Water Res.,* 17: 173.

Horikoshi, S., H. Hidaka, and N. Serpone (2001) Photocatalyzed degradation of polymers in aqueous semiconductor suspsensions V. Photomineralization of Lactam ring-pendant Polyvinylpyrrolidone at Titania/water interfaces, *J. of Photochemistry and Photobiology A: Chemistry,* 138: 69.

Hong, S., C.M. Duttweiler, and A.T. Lemley (1999) Analysis of MTBE and its degradation by-products by direct aqueous injection onto gas chromatography with mass spectrophotometry or flame ionization detection system. *Journal of Chromatography*. A, 857, 205–216.

Huang, F.Y.C. and K.Y. Liu (1999) Treatability studies of groundwater contaminated with Bis(2-chloroethyl) ether. *Environ. Progress*. 18, 55–59.

Ito, K., W. Jian, W. Nishijima, A.U. Baes, E. Shoto, and M. Okada (1998). "Comparison of Ozonation and AOPs combined with biodegradation for removal of THM precursors in treated sewage effluents", *Water Sci. & Technol.*, 38 (7):179–186.

Kaludjerski. M. "Enhancement of Biodegradation of DCDE by Pre-Oxidation" M.S. Thesis. San Diego State University, 2001.

Karpel Vel Leitner, N. Papilhou, A.L. Croue, J.P. Peyrot, and M. Dore (1994). "Oxidation of MTBE and ETBE by ozone and combined ozone/hydrogen peroxide" *Ozone Sci. &Eng.,* 16, (1), 41–54.

Kearney, P.C., M.T. Muldoon, C.J. Somich, J.M. Ruth, and D.J. Voaden (1988) Biodegradation of ozonated atrazine as a wastewater disposal system. *J. Agric. Food Chem.* 36, 1301–1306.

Kearney, P.C., Q. Zeng, and J.M. Ruth (1983) Oxidative pretreatment accelerates TNT metabolism in soils. *Chemosphere*, 12 (11/12): 1583–1597.

Kitis, M., C.D. Adams, and G.T. Daigger (1999) The Effect of Fenton's reagent treatment on the biodegradability of nonionic surfactants: Laboratory studies and economic analysis, *Water Res.*, 33(11): 2561–2568.

Kitis, M., C.D. Adams, J. Kuzhikannil, and G.T. Daiger (2000) Effects of ozone/hydrogen peroxide pretreatment on aerobic biodegradability of nonionic surfactants and polypropylene glycol, *Environ. Sci. Technol.*, 34, 2305–2310.

Kiwi, J., C.P. Pulgarin, P.Peringer, and M. Gratzel (1993) Beneficial effects of homogeneous photo-fenton pretreatment upon the biodegradation of anthraquinone sulfonate in wastewater treatment. *Appl. Catal.B.* 3(1):85–99.

Kiwi, J., A.Lopez, and V. Nadtochenko (2000) Mechanism and kinetics of the OH-radical Intervention during Fenton oxidation in the presence of a significant amount of radical scavenger (Cl^-), *Environ. Sci. Technol.*, 34, 2162–2168.

Kornmüller, A. and U. Weismann (1999) Continuous ozonation of polycyclic aromatic hydrocarbons in oil/water-emulsions and biodegradation of oxidation products, *Water Sci. & Technol.*, 40(4–5)107–114.

Kolthoff , I.M., E.B. Sandell, E.J. Meehan, and S. Bruckenstein, (1969). Quantitative chemical analysis. Macmillan. Toronto, Canada.

Kuo, C., (1998). Improved application of ion chromatographic determination of carboxylic acids in ozonated drinking water. *Journal of Chromatography .A,* 804: 265.

Lee, S.H and J.B. Carberry (1992) Biodegradation of PCP enhanced by chemical oxidation pretreatment. *Water Environ. Res.* 64 (5): 682–690.

Levy, G. and D. Fergus, "Microdetermination of Polyvinylpyrrolidone in Aqueous Solution and in Body Fluids", Analytical Chemistry, 25(9), Sep. 1953.

Lewis, R.J. Sr. (1991). *Hazardous Chemicals Desk Reference.* Van Nostrand Reinhold, 2nd Edition, New York, NY

Li, K.Y., Liu, C.C, and Ni, Q. (1995) Kinetic study of UV peroxidation of bis (2-chloroethyl) ether in aqueous solution. *Ind. Eng. Chem. Res.* 34, 1960–1968.

Liang, S., R.S. Yates, D.V. David, S.J. Pastor, L.S. Palencia, and J.Bruno (2001) Treatibility of MTBE-contaminated groundwater by ozone and peroxide. *Journal AWWA.*, 93 (6): 110–120.

Liao, C.H. and M.D. Gurol (1995) Chemical oxidation by photolytic decomposition of hydrogen peroxide, *Env. Sci. & Tech.*, 29, 12 pp. 3007–3014.

Lin, S.S. and M.D. Gurol (1994) Investigation of H_2O_2 catalysis with iron species, Poster presentation at the Annual Conference of AWWA, New York, N.Y.

Lin, S.S. and M.D. Gurol (1998) Catalytic decomposition of hydrogen peroxide on iron oxide: Kinetics, mechanisms and implications, *Env. Sci. & Tech.*, 32(10)1417.

Lunar, L., D. Sicilia, S. Rubio, D. Perez-Bendito, and U. Nickel (2000) Degradation of photographic developers by Fenton's reagent , *Water Res.* 34:1791–1802.

Manilal, V.B., A. Haridas, R. Alexander, and G.D. Surrender (1992). Photocatalytic treatment of toxic organics in wastewater: Toxicity and photodegradation products. *Water Res.* 26 (8) 1035–1038.

Marco, A., S. Esplugas, and G. Saum (1997). How and why combine chemical and biological processes for wastewater treatment, *Water Sci. & Technol*, 35(4): 321–327.

Mitani, M.M., A.A. Keller, B.A. Clifford, R.G. Rinker, and O.C. Sandall (2002) Kinetics and products of reactions of MTBE with ozone and ozone/hydrogen peroxide in water, *Journal of Hazardous Materials*, 89(2–3): 197–212.

Mo, K., C.O. Lora, A.E. Wanken, M. Javanmardian, X.Yang, and C.F. Kulpa (1997) Biodegradation of methyl *t-butyl* ether by pure bacterial cultures, *App. Microbi.& Technol.* 47: 69–72.

Mohanty, N.R. and I.W. Wei (1993). Oxidation of 2.4-dinitrotoluene using Fenton's reagent: Reaction mechanisms and their practical applications. *Hazardous Waste & Hazardous Materials*, 10 (2), 171–183.

OECD Guidelines for testing of chemicals, OECD, Paris, France, 1987.

Pitter, P. and Chudoba, J., Biodegradability of Organic Substances in the Aquatic Environment, CRC Press, Boca Raton, FL, 1990.

Rahn, R.O., P. Xu, and S.L. Miller (1999) Dosimetry of room-air germicidal (254 nm) radiation using spherical actinometry, *Photochemistry and Photobiology*, 70(3), 314, 1999.

Ravikumar, J.X. and M.D. Gurol (1991) Effectiveness of chemical oxidation in enhancing the biodegradation of pentachlorophenol in soil: A Laboratory study *Proc. 23rd Mid-Atlantic Industrial Waste Confer.* 211–221, Pittsburgh, PA .

Ravikumar, J. (1992) "Treatment of organic contaminants on soil by chemical oxidation and chemical oxidation coupled with biodegradation." Ph.D Dissertation, Drexel University.

Safarzadeh-Amiri, A. (2001) O_3/H_2O_2 treatment of methyl-*tert*-butyl ether (MTBE) in contaminated waters. *Water Research*, Volume 35, Issue 15, 3706–3714.

Sawyer, D.T., A. Sobkoviak, and T. Matsushita. (1996) Metal/hydroperoxide-induced activation of dioxygen for the oxygenation of hydrocarbons: oxygenated Fenton chemistry. *Acc. Chem. Res.* 29, 409–416.

Schwarzenbach, R.P., P.M. Gschwend, and D.M. Imboden (1993), *Environmental Organic Chemistry*, John Wiley & Sons, New York, NY.

Scott, J.P. and D.F. Ollis (1995) Integration of chemical and biological oxidation processes for water treatment: Review and recommendation. *Environ. Progress* 14 (2):88–103.

Somich, C.J., M.T. Muldoon, and P.C. Kearney. (1990) On-site treatment of pesticide waste and rinsate using ozone and biologically active soil. *Environ. Sci. Technol.* 25 (5): 745–749.

Somich , C.J., C. Kearney, M.T. Muldoon, and S. Elsasser (1988). Enhanced soil degradation of alachlor by treatment with ultraviolet light and ozone. *J. Agric. Food Chem.* 36: 1322–1326.

Standard Methods for the Examination of Water and Wastewaters, (19th ed). APHA, AWWA, WEF (1995).

Stefan, M., J. Mack, and R. Bolton (2000) Degradation pathways during the treatment of methyl *tert*-butyl ether by the UV/H2O2 process, *Environ. Sci. Technol,* 34(4): 650–658.

Suzuki, J., H. Nakagawa, and H. Ito (1976) Study on ozone treatment of water-soluble polymers: II.Utilization of ozonized polyethylene glycol by bacteria. *J. Appl. Ploymer Sci.* 20: 2791–2797.

Volskay, V.T., C.P. Grady, Jr., and H.H. Tabak (1990) Effect of selected RCRA compounds on activated sludge activity*, Research Journal WPCF*, 62(5): 654.

Volskay, V.T. and C.P.L. Grady, Jr., (1990) Respiration inhibition kinetic analysis, *Water Res.,* 24(7): 863.

Wagler J. and J. Malley, Jr. (1994) The removal of methyl *tert*-butyl ether from a model groundwater using UV/peroxide oxidation, *J. of the New England Water Works Assoc.,* 108(3): 236–260.

Walling, C. (1975) Fenton's reagent revisited. *Accounts of Chemical Res.* 8, 125–131.

Walling, C. and M. Clearly (1977) Oxygen evaluation as a critical test of mechanism in the ferric-ion catalyzed decomposition of hydrogen peroxide. *Internat. J.* Chemical Kinetics. IX, 95-601.

Yeber, C.M., J. Rodriguez, J. Baeza, J. Freer, C. Zaror, N. Duran, N., and H.D. Mansilla (1999) Toxicity abatement and biodegradability enhancement of pulp mill bleaching effluent by advanced chemical oxidation, *Water Sci. &Technol.* , 40(11–12): 337–342.

Yurteri, C. and M.D. Gurol (1987) Removal of dissolved organic contaminants by ozonation*, Environ. Progress*, 6(4): 240–245.

WASTEWATER UTILITY

Alabama
Montgomery Water Works & Sanitary Sewer Board

Alaska
Anchorage Water & Wastewater Utility

Arizona
Gila Resources

Glendale, City of, Utilities Department

Mesa, City of

Peoria, City of

Pima County Wastewater Management

Arkansas
Little Rock Wastewater Utility

California
Calaveras County Water District

Central Contra Costa Sanitary District

Contra Costa Water District

Crestline Sanitation District

Delta Diablo Sanitation District

Dublin San Ramon Services District

East Bay Dischargers Authority

East Bay Municipal Utility District

El Dorado Irrigation District

Fairfield-Suisun Sewer District

Fresno Department of Public Utilities

Irvine Ranch Water District

Las Virgenes Municipal Water District

Livermore, City of

Lodi, City of

Los Angeles, City of

Los Angeles County, Sanitation Districts of

Napa Sanitation District

Orange County Sanitation District

Palo Alto, City of

Riverside, City of

Sacramento Regional County Sanitation District

San Diego Metropolitan Wastewater Department, City of

San Francisco, City & County of

San Jose, City of

Santa Barbara, City of

Santa Rosa, City of

South Bayside System Authority

South Coast Water District

South Orange County Wastewater Authority

Sunnyvale, City of

Union Sanitary District

West Valley Sanitation District

Colorado
Boulder, City of

Colorado Springs Utilities

Littleton/Englewood Water Pollution Control Plant

Metro Wastewater Reclamation District, Denver

Connecticut
The Mattabasset District

New Haven, City of, WPCA

District of Columbia
District of Columbia Water & Sewer Authority

Florida
Broward, County of

Fort Lauderdale, City of

Gainesville Regional Utilities

JEA

Miami-Dade Water & Sewer Authority

Orange County Utilities Department

Orlando, City of

Reedy Creek Improvement District

Seminole County Environmental Services

St. Petersburg, City of

Stuart Public Utilities

Tallahassee, City of

Tampa, City of

Toho Water Authority

West Palm Beach, City of

Georgia
Atlanta Department of Watershed Management

Augusta, City of

Clayton County Water Authority

Cobb County Water System

Columbus Water Works

Fulton County

Gwinnett County Department of Public Utilities

Savannah, City of

Hawaii
Honolulu, City and County of

Illinois
American Bottoms Wastewater Treatment Plant

Greater Peoria Sanitary District

Kankakee River Metropolitan Agency

Metropolitan Water Reclamation District of Greater Chicago

North Shore Sanitary District

Wheaton Sanitary District

Iowa
Ames, City of

Cedar Rapids Wastewater Facility

Des Moines Metro Wastewater Reclamation Authority

Iowa City, City of

Kansas
Johnson County Unified Wastewater Districts

Unified Government of Wyandotte County/Kansas City, City of

Kentucky
Louisville & Jefferson County Metropolitan Sewer District

Louisiana
Sewerage & Water Board of New Orleans

Maine
Bangor, City of

Portland Water District

Maryland
Anne Arundel County Bureau of Utility Operations

Howard County Department of Public Works

Washington Suburban Sanitary Commission

Massachusetts
Boston Water & Sewer Commission

Upper Blackstone Water Pollution Abatement District

Michigan
Ann Arbor, City of

Detroit, City of

Holland Board of Public Works

Lansing, City of

Owosso Mid-Shiawassee County WWTP

Saginaw, City of

Wayne County Department of Environment

Wyoming, City of

Minnesota
Rochester, City of

Western Lake Superior Sanitary District

Missouri
Independence, City of

Kansas City Missouri Water Services Department

Little Blue Valley Sewer District

Nebraska
Lincoln Wastewater System

Nevada
Henderson, City of

New Jersey
Bergen County Utilities Authority

Passaic Valley Sewerage Commissioners

New York
New York City Department of Environmental Protection

Rockland County Solid Waste Management Authority/ Sewer District

North Carolina
Charlotte/Mecklenburg Utilities

Durham, City of

Metropolitan Sewerage District of Buncombe County

Orange Water & Sewer Authority

Ohio
Akron, City of

Butler County Department of Environmental Services

Columbus, City of

Metropolitan Sewer District of Greater Cincinnati

Northeast Ohio Regional Sewer District

Summit, County of

Oklahoma
Tulsa, City of

Oregon
Clean Water Services

Eugene/Springfield Water Pollution Control

Water Environment Services

Pennsylvania
Philadelphia, City of

University Area Joint Authority, State College

South Carolina
Charleston Commissioners of Public Works

Mount Pleasant Waterworks & Sewer Commission

Spartanburg Sanitary Sewer District

Tennessee
Cleveland, City of

Knoxville Utilities Board

Murfreesboro Water & Sewer Department

Nashville Metro Water Services

Texas
Austin, City of

Dallas Water Utilities

Denton, City of

El Paso Water Utilities

Fort Worth, City of

Gulf Coast Waste Disposal Authority

Houston, City of

San Antonio Water System

Trinity River Authority

Utah
Salt Lake City Corporation

Virginia
Alexandria Sanitation Authority

Arlington, County of

Fairfax County Virginia

Hampton Roads Sanitation District

Henrico, County of

Hopewell Regional Wastewater Treatment Facility

Loudoun County Sanitation Authority

Lynchburg Regional WWTP

Prince William County Service Authority

Richmond, City of

Rivanna Water & Sewer Authority

Washington
Edmonds, City of

Everett, City of

King County Department of Natural Resources

Seattle Public Utilities

Sunnyside, Port of

Yakima, City of

Wisconsin
Green Bay Metro Sewerage District

Madison Metropolitan Sewerage District

Milwaukee Metropolitan Sewerage District

Racine, City of

Sheboygan Regional Wastewater Treatment

Wausau Water Works

Australia
South Australian Water Corp.

Sydney Water Corp.

Water Corp. of Western Australia

Canada
Greater Vancouver Regional District

Toronto, City of, Ontario

Winnipeg, City of, Manitoba

Mexico
Servicios de Agua y Drenaje de Monterrey, I.P.D.

New Zealand
Watercare Services Limited

Singapore
Singapore Public Utilities Board

United Kingdom
Yorkshire Water Services Limited

STORMWATER UTILITY

California
Los Angeles, City of, Department of Public Works

Monterey, City of

San Francisco, City & County of

Santa Rosa, City of

Sunnyvale, City of

Colorado
Boulder, City of

Georgia
Griffin, City of

Iowa
Cedar Rapids Wastewater Facility

Des Moines Metro Wastewater Reclamation Authority

Kansas
Overland Park, City of

Kentucky
Louisville & Jefferson County Metropolitan Sewer District

Maine
Portland Water District

Minnesota
Western Lake Superior Sanitary District

North Carolina
Charlotte, City of, Stormwater Services

Pennsylvania
Philadelphia, City of

Tennessee
Chattanooga Storm Water Management

Washington
Seattle Public Utilities

STATE

Arkansas Department of Environmental Quality

Fresno Metropolitan Flood Control District, Calif.

Urban Drainage & Flood Control District, Colo.

CORPORATE

ADS Environmental Services

The ADVENT Group Inc.

Alan Plummer & Associates

Alden Research Laboratory

Aqua-Aerobic Systems, Inc.

Aquateam–Norwegian Water Technology Centre A/S

BaySaver Inc.

BioVir Laboratories, Inc.

Black & Veatch

Boyle Engineering Corporation

Brown & Caldwell

Burns & McDonnell

CABE Associates Inc.

The Cadmus Group

Camp Dresser & McKee Inc.

Carollo Engineers Inc.

Carpenter Environmental Associates Inc.

CDS Technologies Inc.

Chemtrac Systems Inc.

CH2M HILL

Damon S. Williams Associates, LLC

David L. Sheridan, P.C.

Dewling Associates, Inc.

Earth Tech Inc.

Eco-Matrix

Ecovation

Environmental Engineers International

EMA Inc.

The Eshelman Company Inc.

Finkbeiner, Pettis, & Strout Inc.

Freese & Nichols, Inc.

ftn Associates Inc.

Fuss & O'Neill, Inc.

Gannett Fleming Inc.

Golder Associates Inc.

Greeley and Hansen LLC

Hazen & Sawyer, P.C.

HDR Engineering Inc.

HNTB Corporation

HydroQual Inc.

Infilco Degremont Inc.

Ingersoll-Rand Energy Systems

Insituform Technologies Inc.

Institute for Environmental Technology & Industry, Korea

Jacobson Helgoth Consultants Inc.

Jason Consultants Inc.

Jordan, Jones, & Goulding Inc.

KCI Technologies Inc.

Kelly & Weaver, P.C.

Kennedy/Jenks Consultants

Komline Sanderson Engineering Corporation

Lawler, Matusky & Skelly Engineers, LLP

Limno-Tech Inc.

Lombardo Associates Inc.

Malcolm Pirnie Inc.

McKim & Creed

MEC Analytical Systems Inc.

Metcalf & Eddy Inc.

MWH

New England Organics

Odor & Corrosion Technology Consultants Inc. (OCTC)

Oswald Green, LLC

PA Government Services Inc.

Parametrix Inc.

Parsons

Post, Buckley, Schuh & Jernigan

R&D Engineering/Conestoga Rover & Associates

The RETEC Group

RMC, Inc. (Raines, Melton & Carella)

R.M. Towill Corporation

Ross & Associates Ltd.

Royce Technologies

SAIC Maritime Technical Group

Short Elliott Hendrickson, Inc.

Stantec Consulting Group, Inc.

Stormwater Management, Inc.

Synagro Technologies, Inc.

Tetra Tech Inc.

Trojan Technologies Inc.

URS Corporation

Veolia Water NATC

Wade-Trim Inc.

Weston Solutions Inc.

Woodard & Curran

WRc/D&B, LLC

WWETCO, LLC

Zoeller Pump Company

INDUSTRY

American Electric Power

ChevronTexaco Energy Research & Technology Company

The Coca-Cola Company

Dow Chemical Company

DuPont Company

Eastman Kodak Company

Eli Lilly & Company

Merck & Company Inc.

ONDEO Services

Procter & Gamble Company

PSEG Services Corp.

RWE Thames Water Plc

Severn Trent Services Inc.

United Water Services LLC

Note: List as of 7/1/04

WERF Product Order Form

As a benefit of joining the Water Environment Research Foundation, subscribers are entitled to receive one complimentary copy of all final reports and other products.Additional copies are available at cost (usually $10).To order your complimentary copy of a report, please write "free" in the unit price column.WERF keeps track of all orders.If the charge differs from what is shown here, we will call to confirm the total before processing.

Name Title

Organization

Address

City State Zip Code Country

Phone Fax Email

Stock #	Product	Quantity	Unit Price	Total

Postage & Handling

VA Residents Add 4.5% Sales Tax

Canadian Residents Add 7% GST

TOTAL

Method of Payment:

(All orders must be prepaid.)

❑ Check or Money Order Enclosed

❑ Visa ❑ Mastercard ❑ American Express

Account No. Exp. Date

Signature

Shipping & Handling:

Amount of Order	United States	Canada & Mexico	All Others
Up to but not more than:	Add:	Add:	Add:
$20.00	$5.00*	$8.00	50% of amount
30.00	5.50	8.00	40% of amount
40.00	6.00	8.00	
50.00	6.50	14.00	
60.00	7.00	14.00	
80.00	8.00	14.00	
100.00	10.00	21.00	
150.00	12.50	28.00	
200.00	15.00	35.00	
More than $200.00	Add 20% of order	Add 20% of order	

*minimum amount for all orders

Note: Please make checks payable to the Water Environment Research Foundation.

To Order (Subscribers Only):

Log on to www.werf.org and click on the "Product Catalog."

Phone: (703) 684-2470
Fax: (703) 299-0742.

WERF
Attn: Subscriber Services
635 Slaters Lane
Alexandria, VA 22314-1177

To Order (Non-Subscribers):

Non-subscribers may be able to order WERF publications either through WEF (www.wef.org) or IWAP (www.iwapublishing.com).Visit WERF's website at www.werf.org for details.

IWA Publishing's authorised EU representative for General Product Safety Regulations is Diane D'Arras, 15 rue Duret, 75116 Paris, France, e-mail: safety@iwap.co.uk.

Printed and bound by CPI Group (UK) Ltd, Croydon, CR0 4YY
17/09/2026
02218811-0001